Dwain Marshall

Experiments in Analog Fundamentals

A Systems Approach

Teacher
Mr Johnston

David M. Buchla

PEARSON

Boston Columbus Indianapolis New York San Francisco Upper Saddle River

Amsterdam Cape Town Dubai London Madrid Milan Munich Paris Montreal Toronto

Delhi Mexico City Sao Paulo Sydney Hong Kong Seoul Singapore Taipei Tokyo

Editorial Director: Vernon R. Anthony
Senior Acquisitions Editor: Lindsey Prudhomme
Development Editor: Dan Trudden
Editorial Assistant: Yvette Schlarman
Director of Marketing: David Gesell
Marketing Manager: Harper Coles
Senior Marketing Coordinator: Alicia Wozniak
Senior Marketing Assistant: Les Roberts
Senior Managing Editor: JoEllen Gohr
Senior Project Manager: Rex Davidson
Senior Operations Supervisor: Pat Tonneman
Cover Image: stoupa/Shutterstock.com
Media Project Manager: Karen Bretz
Composition: Naomi Sysak
Printer/Binder: Edwards Brothers Malloy
Cover Printer: Edwards Brothers Malloy

10 9 8 7 6 5 4 3 2

ISBN 10: 0-13-298867-4
ISBN 13: 978-0-13-298867-4

Preface

This laboratory manual is designed to be used with *Analog Fundamentals: A Systems Approach*, by Thomas Floyd and David Buchla. Electronics is traditionally a laboratory course that is meaningful to students only when the theory is reinforced and observed in practice. These experiments are designed to be practical investigations that provide the student with a permanent record of data, results, and answers to questions. There are 38 experiments written with a consistent format that is described below. A cross-listing of the experiments and the relevant text material is presented on page vi.

Experiments are supported with PowerPoint® slides that are provided free of charge to instructors who have adopted this lab manual. Each experiment has a set of slides that are designed as a class review of the experiment by providing a Troubleshooting and Related Problem. PowerPoint® slides are available to instructors at http://pearsonhighered.com/floyd/.

To access supplementary materials online, instructors need to request an instructor access code. Go to **www.pearsonhighered.com/irc**, where you can register for an instructor access code. Within 48 hours after registering, you will receive a confirming e-mail, including an instructor access code. Once you have received your code, go to the site and log on for full instructions on downloading the materials you wish to use.

Multisim® files have been prepared with a number of prebuilt, simulated circuits from the lab manual. A short description of *Multisim®* can be found starting on page 11 of this manual. Of course, no computer simulation can replace the practical benefit of actual laboratory work, but a computer simulation has the capability for students to do *what-if* analysis for circuits and to provide a means of seeing how actual circuits compare to the simulation.

Each of the experiments contains the following parts:

Reading: Reading assignments, which are referenced to the *Analog Fundamentals* text.
Objectives: Statements of what the student should be able to do after completing the experiment.
Summary of Theory: The Summary of Theory is intended to reinforce the important concepts in the text with a review of the main points prior to the laboratory experience. In most cases, specific practical information needed in the experiment is presented.
Materials Needed: A list of the components and small items required but not including the equipment found at a typical lab station. Particular care has been exercised to select materials that are readily available, keeping cost at a minimum.
Procedure: This section contains a relatively structured set of steps for performing the experiment. Needed tables, graphs, and figures are positioned close to the first referenced location to avoid confusion. Laboratory techniques are given in detail.
Conclusion: Space is provided for the student to summarize the key findings from the experiment.
Evaluation and Review Questions: This section contains five questions, which require the student to draw conclusions from the laboratory work and check his or her understanding of the concepts. Troubleshooting questions are frequently presented.

For Further Investigation: This section contains specific suggestions for additional related laboratory work. A number of these lend themselves to a formal laboratory report or they can be used as an enhancement to the experiment.

The first experiment is different from the others in this manual in that the student is not asked to compute parameters for the circuit before building it. This "top down" approach introduces important amplifier parameters (gain, transfer curve, saturation) early in the study of analog circuits (in sync with the text). The amplifier in this experiment is a good circuit for students to simulate in *Multisim*, even before characteristic curves and biasing is presented (in Experiments 6 and 7). They will be able to compare the dc and ac voltages with the lab experiment and observe the onset of saturation.

Each laboratory station should contain a dual variable regulated power supply, a function generator, a multimeter, and a dual-channel oscilloscope. It is useful if the laboratory is equipped to measure capacitors and inductors. In addition, a transistor curve tracer and frequency counter are useful but not required. A list of all required materials for all experiments is given in the Appendix; separate material lists are given for each individual experiment.

This manual is specifically designed to follow the sequence of *Analog Fundamentals* but it can be used with other texts by ignoring the references. As in the text, the lab book does not use arrows with current direction, so the experiments work equally well for schools using electron flow or conventional current flow.

I would like to thank Dan Trudden and Rex Davidson at Pearson Education and Lois Porter, who copyedited the manuscript and made many excellent suggestions along the way. Finally, I express my appreciation of the support from my wife Lorraine, who has sacrificed to make time for my projects.

David Buchla

Contents

Cross-listing to *Analog Fundamentals: A Systems Approach*

Introduction to the Student

Preparing for Laboratory Work

The purpose of experimental work is to help you gain a better understanding of the principles of electronics and to give you experience with instruments and methods used by technicians and electronic engineers. Each experiment requires you to use electronic instruments to measure various quantities. The measured data will be recorded and you will need to interpret the measurements and draw conclusions about your work. The ability to measure, interpret, and communicate results is basic to electronic work.

Preparation before coming to the laboratory is an important part of experimental work. You should prepare in advance for every experiment by reading the *Reading*, *Objectives*, and the *Summary of Theory* sections before coming to class. *The Summary of Theory* is *not* intended to replace the theory presented in the text – it is meant only as a short review to jog your memory of key concepts and to provide some insight to the experiment. You should also look over the *Procedure* for the experiment. This prelab preparation will enable you to work efficiently in the laboratory and enhance the value of the laboratory time.

This laboratory manual is designed to help you measure and record data as efficiently as possible. Blank data tables and plots are provided where necessary. You will need to interpret and discuss the results in the *Conclusion* section and answer the *Evaluation and Review Questions*. The *Conclusion* to an experiment is a concise statement of your key findings from the experiment. Be careful of generalizations that are not supported by the data. The conclusion should be a specific statement that includes important findings with a brief discussion of problems, revisions, or suggestions you may have for improving the circuit. It should directly relate to the objectives of the experiment.

The Laboratory Notebook

Your instructor may assign a formal laboratory report or a report may be assigned in the section titled *For Further Investigation*. A suggested format for formal reports is as follows:

1. *Title and date.*
2. *Purpose:* Give a statement of what you intend to determine as a result of the investigation.
3. *Equipment and materials:* Include a list of equipment model and serial numbers which can allow retracing if a defective or uncalibrated piece of equipment was used.
4. *Procedure:* Give a brief description of what you did and what measurements you made. A diagram or schematic is often useful.
5. *Data:* Tabulate raw (unprocessed) data; data may be presented in graph form.
6. *Sample calculations:* Give the formulas that you applied to the raw data to transform it to processed data.
7. *Conclusion:* The conclusion is a specific statement supported by the experimental data. It should relate to the objectives for the experiment. For example, if the purpose of the experiment was to determine the frequency response of a filter, the conclusion should describe the frequency response or contain a reference to an illustration of the response.

Graphing

A graph is a pictorial representation of data that enables you to see the effect of one variable on another. Graphs are widely used in experimental work to present information because they enable the reader to discern variations in magnitude, slope, and direction between two quantities. In this manual, you will graph data in many experiments. The following steps will guide you in preparing a graph:

1. Determine the type of scale that will be used. A linear scale is the most frequently used and will be discussed here. Choose a scale factor that enables all of the data to be plotted on the graph without being cramped. The most common scales are 1, 2, 5, or 10 units per division. Start both axes from 0 unless the data covers less than half of the length of the coordinate.
2. Number the *major* divisions along each axis. Do not number each small division as it will make the graph appear cluttered. Each division must have equal weight. *Note*: The experimental data is not used to number the divisions.
3. Label each axis to indicate the quantity being measured and the measurement units. Usually, the measurement units are given in parentheses.
4. Plot the data points with a small dot with a small circle around each point. If additional sets of data are plotted, use other distinctive symbols (such as triangles) to identify each set.
5. Draw a smooth line that represents the data trend. It is normal practice to consider data points but to ignore minor variations due to experimental errors. (*Exception*: calibration curves and other discontinuous data are connected "dot-to-dot".)
6. Title the graph, indicating with the title what the graph represents. The completed graph should be self-explanatory.

Safety in the Laboratory

Nearly all of the experiments in this lab book are designed for low voltages to minimize electric shock hazard; however, one should never assume that electric circuits are safe. A current of a few milliamps through the body can be lethal. In addition, electronic laboratories often contain other hazards such as chemicals and power tools. For your safety, you should review laboratory safety rules before beginning your work in the lab. In particular, you should:

1. Avoid contact with *any* voltage source. Turn off power before working on circuits.
2. Remove watches, jewelry, rings, and so forth before working on circuits – even those circuits with low voltages – as burns can occur.
3. Know the location of the emergency power-off switch.
4. Never work alone in the laboratory.
5. Keep a neat work area and handle tools properly. Wear safety goggles or gloves when required.
6. Ensure that line cords are in good condition and grounding pins are not missing or bent. Do not defeat the three-wire ground system.
7. Check that transformers and instruments that are plugged into utility lines are properly fused and have no exposed wiring. If you are not certain about a procedure, check with your instructor before you begin.
8. Report any unsafe condition to your instructor.
9. Be aware of and follow laboratory rules.

The Oscilloscope
Analog and Digital Storage Oscilloscopes

The oscilloscope is the most widely used general-purpose measuring instrument because it allows you see a graph of the voltage as a function of time in a circuit. Many circuits have specific timing requirements or phase relationships that can be readily measured with a two-channel oscilloscope. The voltage to be measured is converted into a visible display, which is presented on a screen.

There are two basic types of oscilloscope – analog and digital. In general, they each have specific characteristics. Analog scopes are the classic "real-time" instruments that show the waveform on a cathode-ray tube (CRT). Digital oscilloscopes are rapidly replacing analog scopes because of their ability to store waveforms and because of measurement automation and many other features such as connections for computers. The storage function is so important that it is usually incorporated in the name as a Digital Storage Oscilloscope (DSO). Some higher-end DSOs can emulate an analog scope in a manner that blurs the distinction between the two types. Tektronix, for example, has a line of scopes called DPOs (Digital Phosphor Oscilloscopes) that can characterize a waveform with intensity gradients like an analog scope and gives the benefits of a digital oscilloscope for measurement automation.

Analog and digital scopes have similar functions, and the basic controls are essentially the same for both types (certain enhanced features are not the same). In the descriptions that follow, the analog scope is introduced first to familiarize you with basic controls, then a specific digital storage oscilloscope is described.

Analog Oscilloscopes

Block Diagram

The analog oscilloscope contains four functional blocks, as illustrated in Figure I-1. Shown within these blocks are the most important typical controls found on nearly all oscilloscopes.

Each of two input channels is connected to the vertical section, which can be set to attenuate or amplify the input signals to provide the proper voltage level to the vertical deflection plates of the CRT. In a dual-trace oscilloscope (the most common type), an electronic switch rapidly switches between channels to send one or the other to the display section.

The trigger section samples the input waveform and sends a synchronizing trigger signal at the proper time to the horizontal section. The trigger occurs at the same relative time, thus superimposing each succeeding trace on the previous trace. This action causes the signal to appear to stop, allowing you to examine it.

The horizontal section contains the time-base (or sweep) generator, which produces a linear ramp, or "sweep," waveform that controls the rate the beam moves across the screen. The horizontal position of the beam is proportional to the time that elapsed from the start of the sweep, allowing the horizontal axis to be calibrated in units of time. The output of the horizontal section is applied to the horizontal deflection plates of the CRT.

Finally, the display section contains the CRT and beam controls. It enables the user to obtain a sharp presentation with the proper intensity. The display section usually contains other features such as a probe compensation jack and a beam finder.

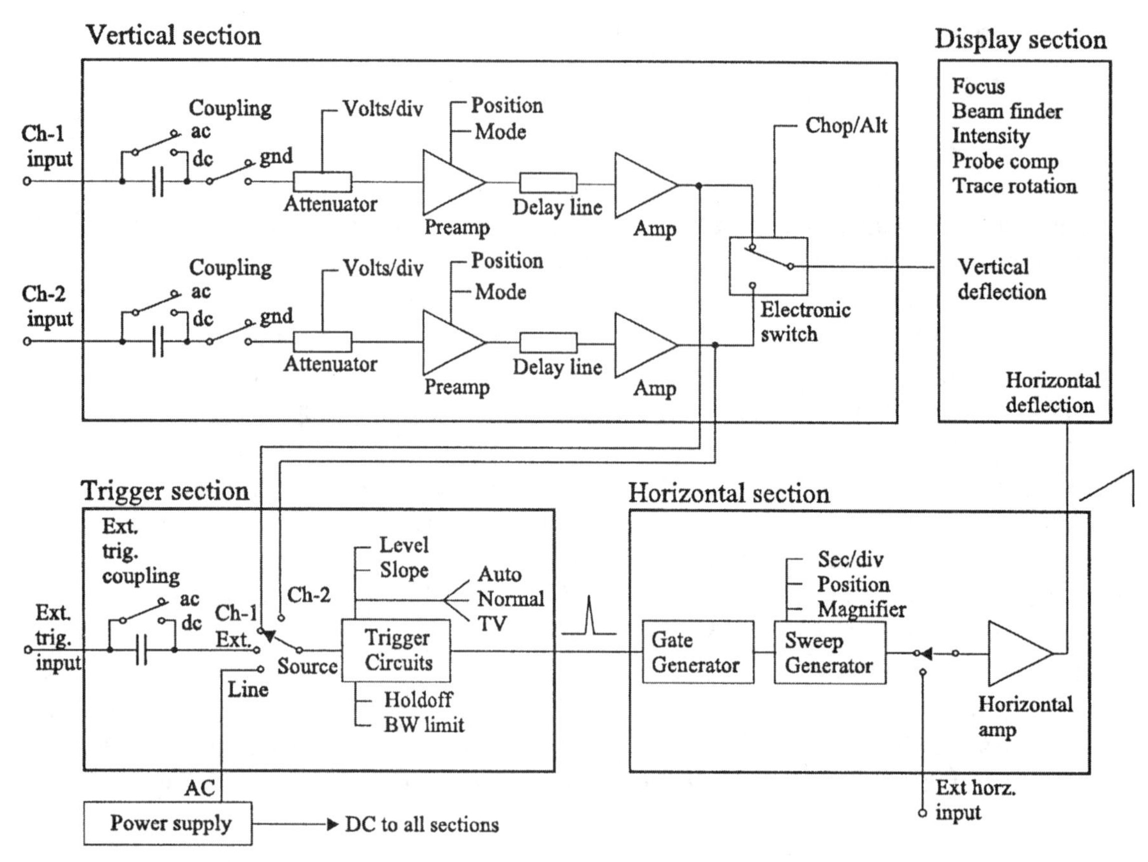

Figure I-1 Block diagram of a basic analog oscilloscope

Controls
Generally controls for each section of the oscilloscope are grouped together according to function. Frequently, there are color clues to help you identify groups of controls. Details of these controls are explained in the operator's manual for the oscilloscope; however, a brief description of frequently used controls is given in the following paragraphs.

The important controls are shown on the block diagram of Figure I-1.

Display Controls The display system contains controls for adjusting the electron beam, including FOCUS and INTENSITY controls. FOCUS and INTENSITY are adjusted for a comfortable viewing level with a sharp focus. The display section may also contain the BEAM FINDER, a control used in combination with the horizontal and vertical POSITION controls to bring the trace on the screen. Another control over the beam intensity is the z-axis input. A control voltage on the z-axis input can be used to turn the beam on or off or adjust its brightness. Some oscilloscopes also include the TRACE ROTATION control in the display section. TRACE ROTATION is used to align the sweep with a horizontal graticule line. This control is usually adjusted with a screwdriver to avoid accidental adjustment. Usually a PROBE COMP connection point is included in the display group of controls. Its purpose is to allow a quick qualitative check on the frequency response of the probe-scope system.

Vertical Controls The vertical controls include the VOLTS/DIV (vertical sensitivity) control and its vernier, the input COUPLING switch, and the vertical POSITION control. There is a duplicate set of these controls for each channel and various switches for selecting channels or other vertical operating modes. The vertical inputs is connected through a selectable attenuator to a high input impedance dc amplifier. The VOLTS/DIV control on each channel selects a combination of attenuation/gain to determine the vertical sensitivity. For example, a low-level signal will need more gain and less attenuation than a higher level signal. The vertical sensitivity is adjusted in fixed VOLTS/DIV increments to allow the user to make calibrated voltage measurements. In addition, a concentric vernier control is usually provided to allow a continuous range of sensitivity. This knob must be in the detent (calibrated) position to make voltage measurements. The detent position can be felt by the user as the knob is turned because the knob tends to "lock" in the detent position. Some oscilloscopes have a warning light or message when the vernier is not in its detent position.

The input coupling switch is a multiple-position switch than can be set for AC, GND, or DC and sometimes includes a 50 Ω position. The GND position of the switch internally disconnects the signal from the scope and grounds the input amplifier. This position is useful if you want to set a ground reference level on the screen for measuring the dc component of a waveform. The AC and DC positions are high-impedance inputs, typically 1 MΩ shunted by 15 pF of capacitance. High-impedance inputs are useful for general probing at frequencies below about 1 MHz. At higher frequencies, the shunt capacitance can load the signal source excessively, causing measurement error. Attenuating divider probes are good for high-frequency probing because they have very high impedance (typically 10 MQ) with very low shunt capacitance (as low as 2.5 pF).

The AC position of the coupling switch inserts a series capacitor before the input attenuator, causing dc components of the signal to be blocked. This position is useful if you want to measure a small ac signal riding on top of a large dc signal-power supply ripple, for example. The DC position is used when you want to view both the AC and DC components of a signal. This position is best when viewing digital signals, because the input *RC* circuit forms a differentiating network. The AC position can distort the digital waveform because of this differentiating circuit. The 50 Ω position places an accurate 50 Ω load to ground. This position provides the proper termination for probing in 50 Ω systems and reduces the effect of a variable load, which can occur in high-impedance termination. The effect of source loading must be taken into account when using a 50 Ω input. It is important not to overload the 50 Ω input, because the resistor is normally rated for only 2 W, implying a maximum of 10 V of signal-can be applied to the input.

The vertical POSITION control varies the dc voltage on the vertical deflection plates, allowing you to position the trace anywhere on the screen. Each channel has its own vertical POSITION control, enabling you to separate the two channels on the screen. You can use vertical POSITION when the coupling switch is in the GND position to set an arbitrary level on the screen as ground reference.

There are two types of dual-channel oscilloscope-dual beam and dual trace. A dual-beam oscilloscope has two independent beams in the CRT and independent vertical deflection systems, allowing both signals to be viewed at the same time. A dual-trace oscilloscope has only one beam and one deflection system, it uses electronic switching to show the two signals. Dual-beam oscilloscopes are generally restricted to high-performance research instruments and are much more expensive than dual-trace oscilloscopes. The block diagram in Figure I-1, is for a typical dual-trace oscilloscope.

A dual-trace oscilloscope has user controls labeled CHOP or ALTERNATE to switch the beam between the channels so that the signals appear to occur simultaneously. The CHOP mode rapidly switches the beam between the two channels at a fixed high speed rate, so the two channels appear to be displayed at the same time. The ALTERNATE mode first completes the sweep for one of the channels and then displays the other channel on the next (or alternate) sweep. When viewing slow signals, the CHOP mode is best because it reduces the flicker that would otherwise be observed. High-speed signals can usually be observed best in ALTERNATE mode to avoid seeing the chop frequency.

Another feature on most dual-trace oscilloscopes is the ability to show the algebraic sum and difference of the two channels. For most measurements, you should have the vertical sensitivity (VOLTS/DIV) on the same setting for both channels. You can use the algebraic sum if you want to compare the balance on push-pull amplifiers, for example. Each amplifier should have identical out-of-phase signals. When the signals are added, the resulting display should be a straight line, indicating balance. You can use the algebraic difference when you want to measure the waveform across an ungrounded component. The probes are connected across the ungrounded component with probe ground connected to circuit ground. Again, the vertical sensitivity (VOLTS/DIV) setting should be the same for each channel. The display will show the algebraic difference of the two signals. The algebraic difference mode also allows you to cancel any unwanted signal that is equal in amplitude and phase and is common to both channels.

Dual-trace oscilloscopes also have an X-Y mode, which causes one of the channels to be graphed on the X-axis and the other channel to be graphed on the Y-axis. This is necessary if you want to change the oscilloscope base line to represent a quantity other than time. Applications include viewing a transfer characteristic (output voltage as a function of input voltage), swept frequency measurements, or showing Lissajous figures for phase measurements. Lissajous figures are patterns formed when sinusoidal waves drive both channels.

Horizontal Controls The horizontal controls include the SEC/DIV control and its vernier, the horizontal magnifier, and the horizontal POSITION control. In addition, the horizontal section may include delayed sweep controls. The SEC/DIV control sets the sweep speed, which controls how fast the electron beam is moved across the screen. The control has a number of calibrated positions divided into steps of 1-2-5 multiples, which allow you to set the exact time interval at which you view the input signal. For example, if the graticule has 10 horizontal divisions and the SEC/DIV control is set to 1.0 ms/div, then the screen will show a total time of 10 ms. The SEC/DIV control usually has a concentric vernier control that allows you to adjust the sweep speed continuously between the calibrated steps. This control must be in the detent position in order to make calibrated time measurements. Many scopes are also equipped with a horizontal magnifier that affects the time base. The magnifier increases the sweep time by the magnification factor, giving you increased resolution of signal details. Any portion of the original sweep can be viewed using the horizontal POSITION control in conjunction with the magnifier. This control actually speeds the sweep time by the magnification factor and therefore affects the calibration of the time base set on the SEC/DIV control. For example, if you are using a 10X magnifier, the SEC/DIV dial setting must be divided by 10.

Trigger Controls The trigger section is the source of most difficulties when learning to operate an oscilloscope. These controls determine the proper time for the sweep to begin in order to produce a stable display. The trigger controls include the MODE switch, SOURCE switch, trigger LEVEL, SLOPE, COUPLING, and variable HOLDOFF controls. In addition, the trigger section includes a connector for applying an EXTERNAL trigger to start the sweep. Trigger controls may also include HIGH or LOW FREQUENCY REJECT Switches and BANDWIDTH LIMITING.

The MODE switch is a multiple-position switch that selects either AUTO or NORMAL (sometimes called TRIGGERED) and may have other positions, such as SINGLE sweep. In the AUTO position, the trigger generator selects an internal oscillator that will trigger the sweep generator as long as no other trigger is available. This mode ensures that a sweep will occur even in the absence of a signal, because the trigger circuits will "free run" in this mode. This allows you to obtain a baseline for adjusting ground reference level or for adjusting the display controls. In the NORMAL or TRIGGERED mode, a trigger is generated from one of three sources selected by the SOURCE switch–the INTERNAL signal, an EXTERNAL trigger source, or the AC LINE. If you are using the internal signal to obtain a trigger, the normal mode will provide a trigger only if a signal is present and other trigger conditions (level, slope) are met. This mode is more versatile than AUTO as it can provide stable triggering for very low to very high frequency signals. The SINGLE position is used primarily for photographing the display.

The trigger LEVEL and SLOPE controls are used to select a specific point on either the rising or failing edge of the input signal for generating a trigger. The trigger SLOPE control determines which edge will generate a trigger, whereas the LEVEL control allows the user to determine the voltage level on the input signal that will start the sweep circuits.

The SOURCE switch selects the trigger source–either from the CH-1 signal, the CH-2 signal, an EXTERNAL trigger source, or the AC LINE. In the CH-1 position, a sample of the signal from channel-1 is used to start the sweep. In the EXTERNAL position, a time-related external signal is used for triggering. The external trigger can be coupled with either AC or DC COUPLING. Thc trigger signal can be coupled with AC COUPLING if the trigger signal is riding on a dc voltage. DC COUPLING is used if the triggers occur at a frequency of less than about 20 Hz. The LINE position causes the trigger to be derived from the ac power source. This synchronizes the sweep with signals that are related to the power line frequency.

The variable HOLDOFF control allows you to exclude otherwise valid triggers until the holdoff time has elapsed. For some signals, particularly complex waveforms or digital pulse trains, obtaining a stable trigger can be a problem. This can occur when one or more valid trigger points occur before the signal repetition time. If every event that the trigger circuits qualified as a trigger were allowed to start a sweep, the display could appear to be unsynchronized. By adjusting the variable HOLDOFF control, the trigger point can be made to coincide with the signal-repetition point.

Oscilloscope Probes

Signals should always be coupled into an oscilloscope through a probe. A probe is used to pick off a signal and couple it to the input with a minimum loading effect on the circuit under test. Various types of probes are provided by manufacturers but the most common type is a 10:1 attenuating probe that is shipped with most general-purpose oscilloscopes. These probes have a short ground lead that should be connected to a nearby circuit ground point to avoid oscillation and power line interference. The ground lead makes a mechanical connection to the test circuit and passes the signal through a flexible, shielded cable to the oscilloscope. The shielding helps protect the signal from external noise pickup.

Begin any session with the oscilloscope by checking the probe compensation on each channel. Adjust the probe for a flat-topped square wave while observing the scope's calibrator output. This is a good signal to check the focus and intensity and verify trace alignment. Check the front-panel controls for the type of measurement you are going to make. Normally, the variable controls (VOLTS/DIV and SEC/DIV) should be in the calibrated (detent) position. The vertical coupling switch is usually placed in the DC position unless the waveform you are interested in has a large dc offset. Trigger holdoff should be in the minimum position unless it is necessary to delay the trigger to obtain a stable sweep.

Digital Storage Oscilloscopes

Block Diagram

The digital storage oscilloscope (DSO) uses a fast analog to digital converter (ADC) on each channel (typically two or four channels) to convert the input voltage into numbers that can be stored in a memory. The digitizer samples the input at a uniform rate called the sample rate; the optimum sample rate depends on the speed of the signal. The process of digitizing the waveform has many advantages for accuracy, triggering, viewing hard to see events, and for waveform analysis. Although the method of acquiring and displaying the waveform is quite different than analog scopes, the basic controls on the instrument are similar.

A block diagram of the basic DSO is shown in Figure I-2. As you can see, functionally, the block diagram is like that of the analog scope. As in the analog oscilloscope, the vertical and horizontal controls include position and sensitivity, which are used to set up the display for the proper scaling.

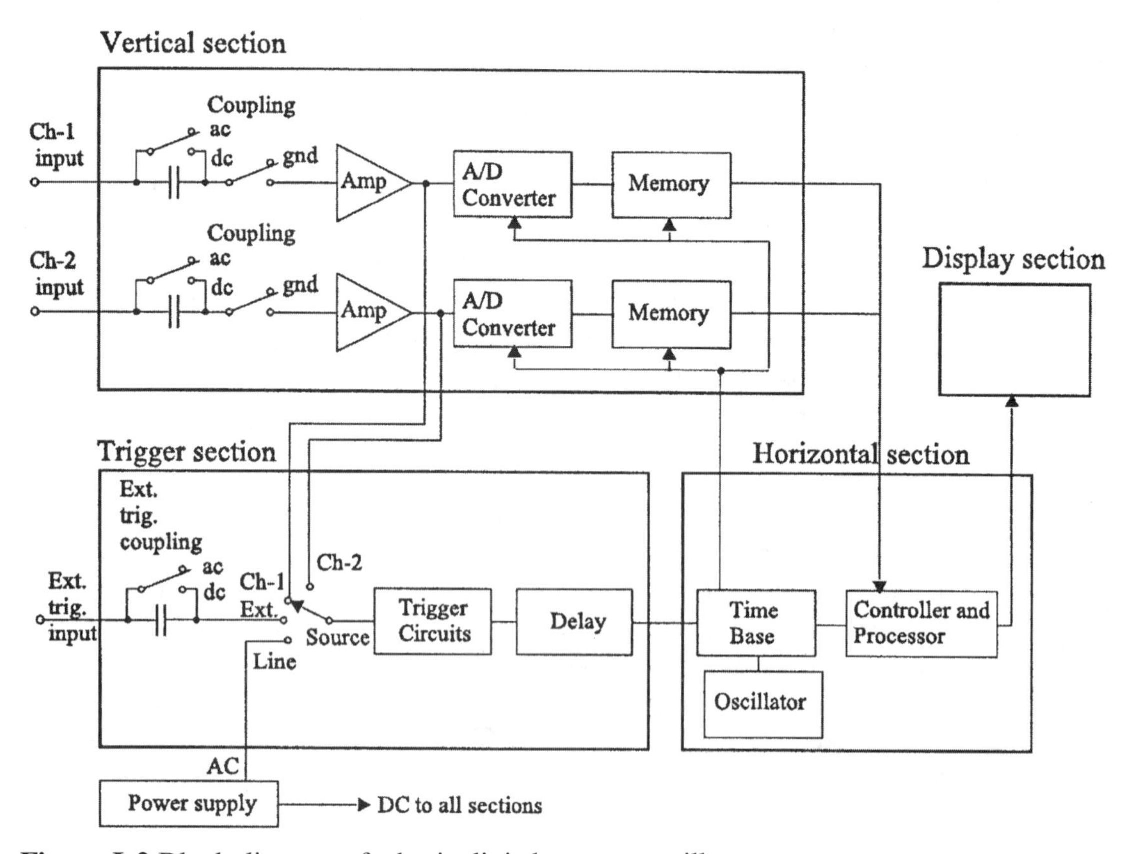

Figure I-2 Block diagram of a basic digital storage oscilloscope

Specifications Important parameters with DSOs include the resolution, maximum digitizing rate, and the size of the acquisition memory as well as the available analysis options. The resolution is determined by the number of bits digitized by the ADC. A low resolution DSO may use only six bits (one part in 64). A typical DSO may use 8 bits, with each channel sampled simultaneously. High end DSOs may use 12 bits. The maximum digitizing rate is important to capture rapidly changing signals; typically the maximum rate is 1 Gs/s. The size of the memory determines the length of time the sample can be taken; it is also important in certain waveform measurement functions.

Triggering One useful feature of digital storage oscilloscopes is their ability to capture waveforms either before or after the trigger event. Any segment of the waveform, either before or after the trigger event can be captured for analysis. **Pretrigger capture** refers to acquisition of data that occurs *before* a trigger event. This is possible because the data is digitized continuously, and a trigger event can be selected to stop the data collection at some point in the sample window. With pretrigger capture, the scope can be triggered on the fault condition, and the signals that preceded the fault condition can be observed. For example, troubleshooting an occasional glitch in a system is one of the most difficult troubleshooting jobs; by employing pretrigger capture, trouble leading to the fault can be analyzed. A similar application of pretrigger capture is in material failure studies where the events leading to failure are most interesting but the failure itself causes the scope triggering.

Beside pretrigger capture, post triggering can also be set to capture data that occur some time after a trigger event. The record that is acquired can begin after the trigger event by some amount of time or by a specific number of events as determined by a counter. A low level response to a strong stimulus signal is an example of when post triggering is useful.

Introduction to Computer Simulation Software and *Multisim*

Computer Simulation Software

Multisim is a computer program that can simulate electronic circuits using a graphical interface to build schematics and test the circuit with simulated instruments. The original computer simulation program was called SPICE (Simulation Program with Integrated Circuit Emphasis), which was developed at the University of California as a computer aid for designing integrated circuits. Today, SPICE is still used as the basic analysis method but with many improvements such as the graphical interface and the ability to convert the circuit to a printed circuit layout.

Although SPICE-based computer simulations are useful for training and as a design aid, they do not replace careful laboratory work. Simulations work well for typical situations, but in certain cases, they do not replicate circuit behavior exactly. From time to time the user will simulate a circuit file and then find that the behavior deviates somewhat from the actual circuit constructed in the laboratory. This is because simulators work with models of circuits that attempt to predict how an actual circuit might react to specific conditions. How well the simulation fits the behavior of real-world circuits depends both upon how accurate the circuit model is and how well the simulation engine works. In several of the circuits in this laboratory manual, there are some known differences between the laboratory experiment and the simulation results.

One example of a difference between simulation software and laboratory work is in the case of oscillators. The imbalances and noise that generally start oscillation in the real world are not present in the simulator, so some means of initiating oscillation must be incorporated. In oscillator circuits which are not capable of self-starting, a switch to force the circuit out of steady state must be included in the simulation. Other cases where the actual and simulated results are different can be due to variation in component parameters, stray capacitance and other parasitic effects, noise pickup, thermal drift, and so forth. The important point is to use simulation as a tool, rather than a replacement for laboratory work.

Multisim

Multisim is a widely used simulation program that calculates the performance of both analog and digital circuits on a computer. Multisim uses a SPICE simulator that provides you with the components and instruments necessary to create board-level designs on your computer, without using breadboards, real components, or actual instruments. Multisim has complete mixed analog and digital simulation and graphical waveform analysis, which allow you to design your circuit and then analyze it using different simulated instruments and analysis options. A brief overview of Multisim is given here.

The Multisim User Interface

The figure on the next page illustrates the user interface. It consists of the menus, toolbars, various windows and a status line.

- **Menus** are a list of commands you use to perform functions and shown across the top. Many of these will be familiar to you if you use other windows based programs. They include the **File Menu, Edit Menu, View Menu,** and so forth. When selected, the menus show a drop down list of available commands.

- **Toolbars** are pictorial buttons you use to modify some aspects of the circuit. Toolbars are grouped to include similar tools and may expand to show additional capabilities. For example, the **Component toolbar** contains buttons that each open a list of related parts, such as basic components, transistors, or integrated circuits and so forth. You can use familiar drag and drop techniques to move components to your circuit. Various toolbars can be displayed by selecting them in the **View Menu**.
- **Windows** are areas that can allow input operations or show results of processes. The **circuit window** is where you create circuit schematics. The **description window** contains text describing the circuit. The **instrument window** shows instrument controls. The **graph window** (not illustrated) shows the results of analyses.
- The **status line** shows the name of the component or instrument to which the cursor is pointing. During simulations, the status line along the bottom indicates the current state of the simulation and the time at which the simulation reached that state. Note that the time is simulation time, not elapsed time, and can be quite different from actual time (called real time).

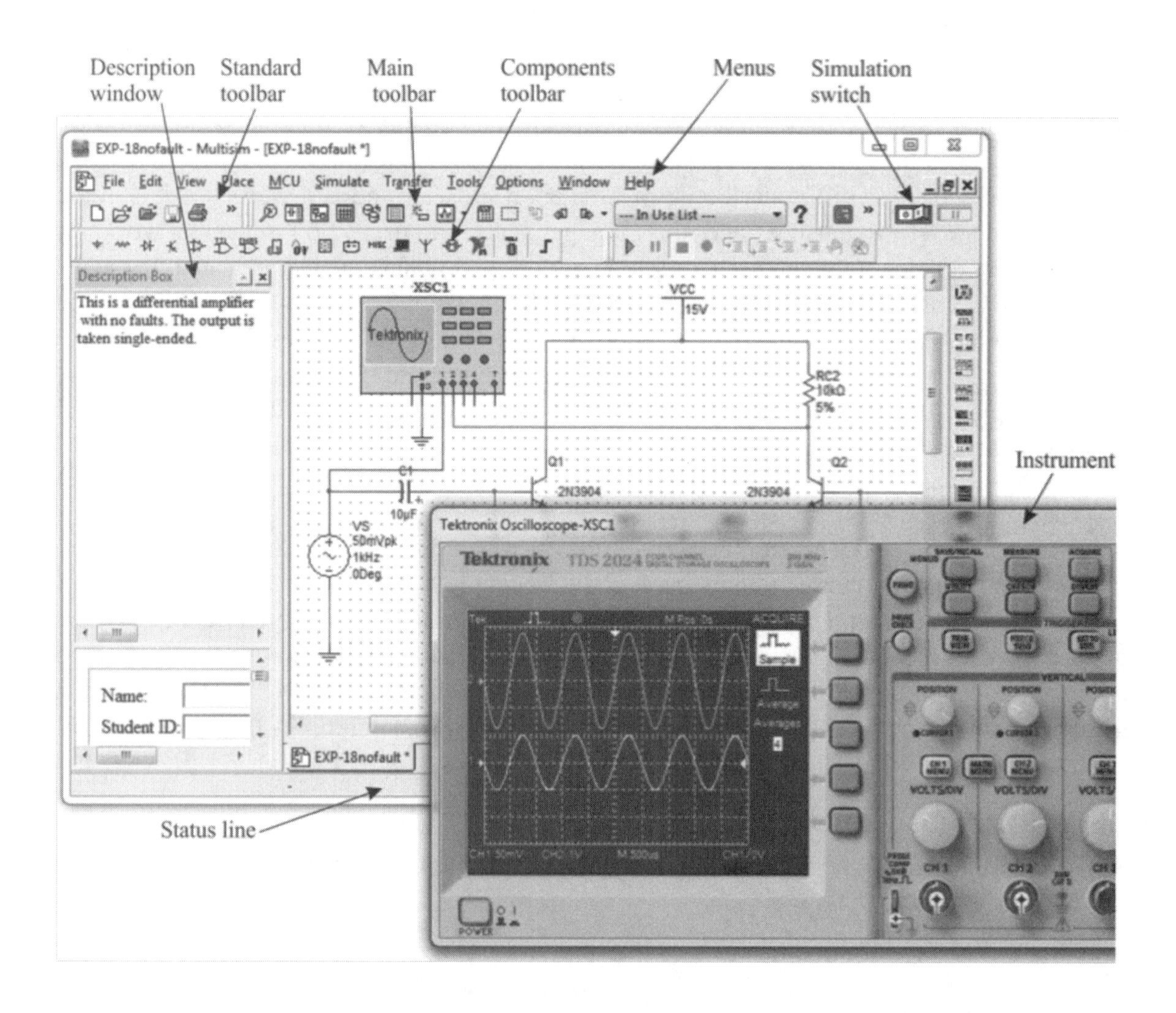

Name ____________

Date ____________

Class ____________

1 Signal Sources and Amplifier Characteristics

Reading:
Floyd and Buchla, *Analog Fundamentals: A Systems Approach*, Sections 1-1 through 1-5

Objectives:
After performing this experiment, you will be able to:

1. Measure the Thevenin resistance of your function generator.
2. Determine the gain of a transistor amplifier for a variety of input voltages.
3. Plot the transfer curve for the amplifier in objective 2.

Summary of Theory:
Thevenin's theorem allows you to replace a two-terminal, linear network with a single source and a series resistor as shown in Figure 1-1. The Thevenin source can be either a dc or an ac source and a series resistance that represents the output resistance of an amplifier or other circuit. The function generator at your lab station is a good example of a complicated circuit that can be replaced with a Thevenin circuit. This enables you to simplify the analysis when a load is connected to the generator, yet still make accurate predictions of the effect. In this experiment, you will begin by determining the equivalent Thevenin circuit of the generator.

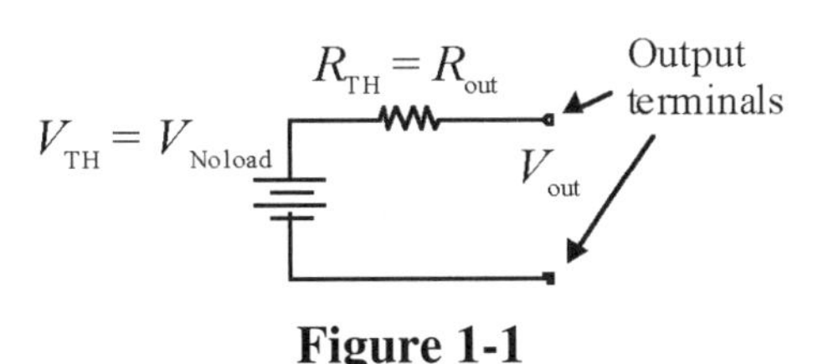

Figure 1-1

Next you will construct and test a low-gain audio amplifier using an *npn* transistor. It isn't necessary to know how to calculate the detailed dc and ac parameters for the amplifier at this point; that will be covered later (in Chapter 3 and Experiments 7, 8, and 9). The object here is to measure the transfer curve (a plot of the output voltage as a function of the input voltage) and to observe saturation as the amplifier is "pushed" beyond its normal operating range. The transfer curve clearly indicates the normal linear range of the amplifier. Transfer curves are used in many applications, including analog to digital converters, digital circuits, and systems with various types of inputs to describe the system behavior.

Materials Needed:
One 2N3904 *npn* transistor (or equivalent)
Resistors: one 1.0 kΩ, one 1.8 kΩ, one 3.9 kΩ, one 10 kΩ
One 10 μF capacitor
One 1.0 kΩ potentiometer

Procedure:

Part 1 - Thevenin Equivalent Circuit for the Function Generator

1. In this part, you will determine the Thevenin equivalent circuit of your function generator. Begin by selecting a 1.0 kHz sinusoidal wave form with no dc offset. Set the amplitude to 200 mV_{pp} with the oscilloscope. This represents the no load voltage from the function generator, so it is the Thevenin voltage for this setting.

2. In this step, you will determine the Thevenin resistance of the generator. This resistance is a fixed quantity that cannot be measured directly. To determine it, a load resistor is connected to the circuit. The Thevenin voltage is now divided between the internal Thevenin resistance and the load. The voltage drop across each resistance is proportional to the resistance, hence the internal resistance can be found by proportion. Most function generators are either 600 Ω or 50 Ω.

   Connect a 1.0 kΩ potentiometer as a load across the output terminals of the function generator as shown in Figure 1-2. Observe the waveform across the potentiometer and adjust it until the waveform is exactly one-half the Thevenin voltage (100 mV_{pp} across the load). At this setting the Thevenin resistance and the load resistance must be equal. Remove the potentiometer from the circuit and measure the resistance of the load and record it.

R_L = ____________________

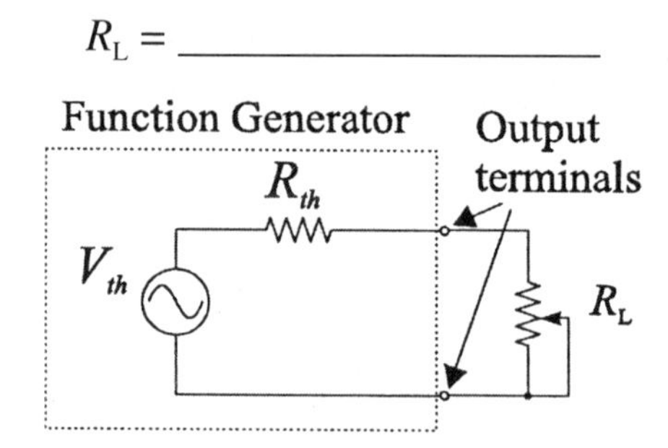

Figure 1-2 Function generator with load

3. Complete the circuit drawing in Figure 1-2 by showing the values for the equivalent Thevenin circuit for your function generator as set in step 1 and measured in step 2.

Part 2 - Transfer Curve for an Amplifier

4. In this part you will measure the transfer curve for the transistor amplifier in Figure 1-3. Measure and record the resistors listed in Table 1-1.

Table 1-1

Resistor	Listed Value	Measured Value
R_1	10 kΩ	
R_2	1.8 kΩ	
R_E	1.0 kΩ	
R_C	3.9 kΩ	

5. Construct the circuit shown in Figure 1-3. The transistor has three leads labeled emitter, base, and collector (reading left to right on the 2N3904). The capacitor isolates transistor bias (dc) from the ac signal. Set the power supply for +15 V. As a quick check that the circuit is wired correctly, check the dc voltage on the emitter and the collector. You should see approximately 1.6 V on the emitter and 8.6 V on the collector. If you do not, recheck the wiring and components.

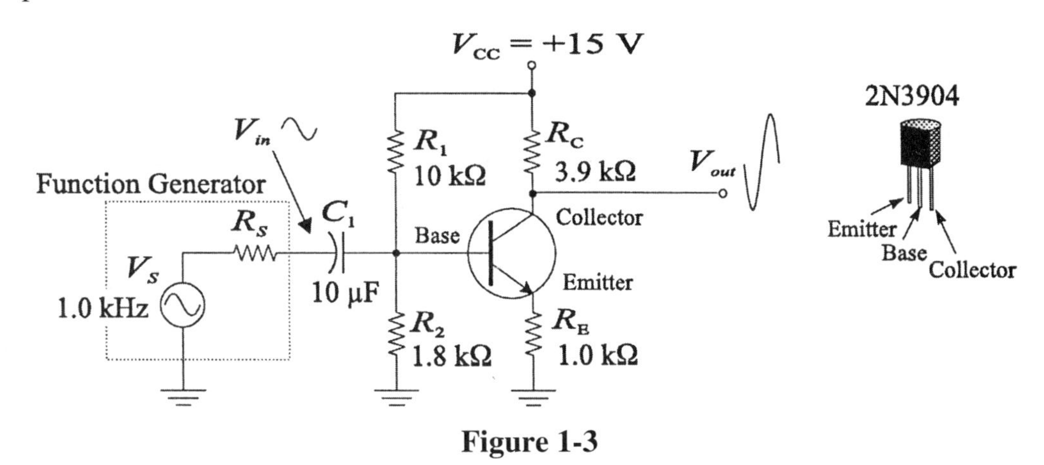

Figure 1-3

6. Using the oscilloscope, set the function generator for a 500 mV_{pp} sinusoidal wave at 1.0 kHz. The generator should be connected to the circuit when setting the voltage. Connect the second channel of the oscilloscope to the output. Measure the peak to peak output voltage and record it in the first line of Table 1-2.

7. Compute the voltage gain, A_v, for a 500 mV_{pp} input by finding the ratio of V_{out} to V_{in}. Record the voltage gain in Table 1-2.

Table 1-2

V_{in}	V_{out}	A_v
500 mV_{pp}		
1.0 V_{pp}		
1.5 V_{pp}		
2.0 V_{pp}		
2.5 V_{pp}		
3.0 V_{pp}		
3.5 V_{pp}		

8. Set the function generator to the next input voltage in Table 1-2. Measure V_{out} and compute the voltage gain. Continue like this for each value in Table 1-2. You will see the amplifier begin to clip the output waveform when the input voltage is about 3.0 V_{pp}. Since a clipped output represents distortion, this is beyond the normal operating region, but you can observe the saturation and the effect on the transfer curve.

9. Plot the transfer curve for the amplifier in Plot 1-1. The transfer curve is an important characteristic for an amplifier and represents the output as a function of the input. Notice that the input voltage is plotted along the *x*-axis because it represents the independent variable; the output voltage is plotted along the *y*-axis because it represents the dependent variable.

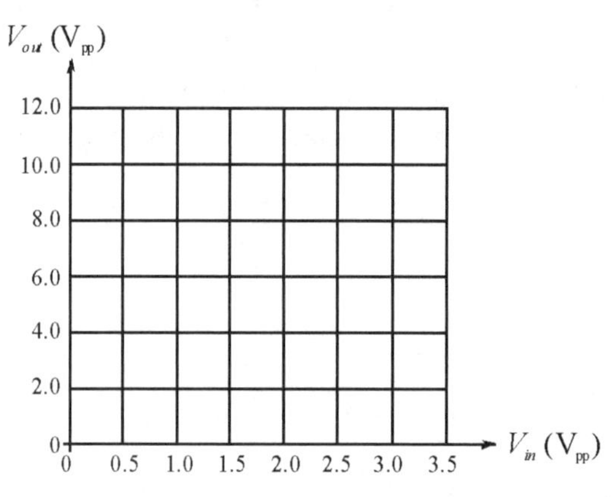

Plot 1-1

Conclusion:

Evaluation and Review Questions:

1. Norton's theorem provides an equivalent circuit for a two-terminal linear source. A Norton circuit has a current source and a parallel resistance (see text, page 16). Convert the Thevenin equivalent for your function generator in Figure 1-2 to its Norton equivalent. Draw the circuit in the space below.

2. Explain why an ohmmeter cannot be used to measure the internal resistance of the function generator.

3. Assume a fixed load of 100 Ω is placed across the function generator used in this experiment. What fraction of the unloaded generator voltage will appear across the load?

4. An amplifier with automatic gain control (AGC) has less gain when the input voltage is high than when it is low. Sketch the general shape of the transfer curve for an AGC amplifier.

5. Decibel voltage gain is given by the equation $A'_v = 20 \log \left(\frac{V_{out}}{V_{in}}\right)$. Compute the decibel voltage voltage gain for the amplifier in part 2 when the input voltage was 1.0 V_{pp}.

For Further Investigation:

The input resistance of the amplifier in Figure 1-3 can only be determined indirectly in a manner similar to finding the internal resistance of the function generator. This time you know the generator resistance (from step 2) but don't know the load resistance represented by the amplifier. With the generator removed from the circuit, set the output of the generator to 1.0 Vrms. Then connect the generator to the amplifier and record the loaded input voltage. Can you use this information to determine the input resistance? Record the input resistance and summarize briefly how you determined it.

Name ______________
Date ______________
Class ______________

2 The Diode Characteristic

Reading:
Floyd and Buchla, *Analog Fundamentals: A Systems Approach*, Sections 2-1 through 2-4

Objectives:
After performing this experiment, you will be able to:
1. Measure and plot the forward- and reverse-biased *I-V* characteristics for a diode.
2. Perform a diode test with an ohmmeter.
3. Test the effect of heat on a diode's response.
4. Measure the ac resistance of a diode.

Summary of Theory:
When a *p*-type material and an *n*-type material are made on the same crystal base, a diode is formed from the *pn* junction. A *pn* junction has unique electrical characteristics. When it is formed, electrons and holes diffuse across the junction, creating a barrier potential which prevents further current without an external voltage source. If a dc voltage source is connected to the diode, the direction it is connected has the effect of either increasing or decreasing the barrier potential. The effect is to allow the diode to either conduct readily in one direction but not the other. If the negative terminal of the source is connected to the *n*-type material and the positive terminal is connected to *p*-type material, the diode is said to be forward-biased and it conducts. If the positive terminal of the source is connected to the *n*-type material and the negative terminal is connected to *p*-type material, the diode is said to be reverse-biased and the diode is a poor conductor.

The schematic symbol for a diode is shown in Figure 2-1. The arrow shows the direction of conventional current (positive to negative). For many applications, thinking of a diode as a one-way valve is sufficient. Other applications require a better approximation.

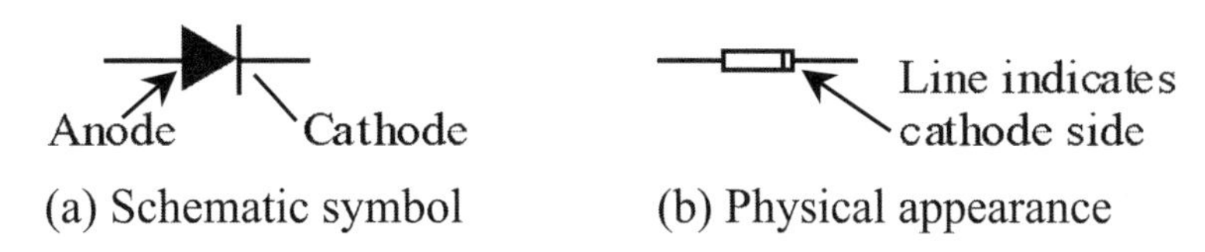

(a) Schematic symbol (b) Physical appearance

Figure 2-1

Diodes can be simplified with three basic models (described briefly here and in Section 2-4 of the text). The ideal model considers the diode as a one-way valve or as an open or closed switch. If it is forward-biased, the switch is closed; if it is reverse-biased the switch is open. The second, practical diode model, adds the forward-biased "diode drop" needed to overcome the barrier potential. For a silicon diode, this is approximately 0.7 V; for germanium, the drop is approximately 0.3 V. The third model adds the ac (or dynamic) resistance of the diode. The ac resistance is found by dividing a small change in voltage by a small change in current. Since the diode is a nonlinear device, this resistance is not constant but will depend on the location of the measured point on the characteristic curve. Figure 2-2 shows a small portion of an *I-V* characteristic curve and illustrates how ac resistance is determined. Note that a small change in voltage is divided by a small change in current to find the ac resistance at the point where $V_D = 0.5$ V. This is the reciprocal of the slope at the point to be measured.

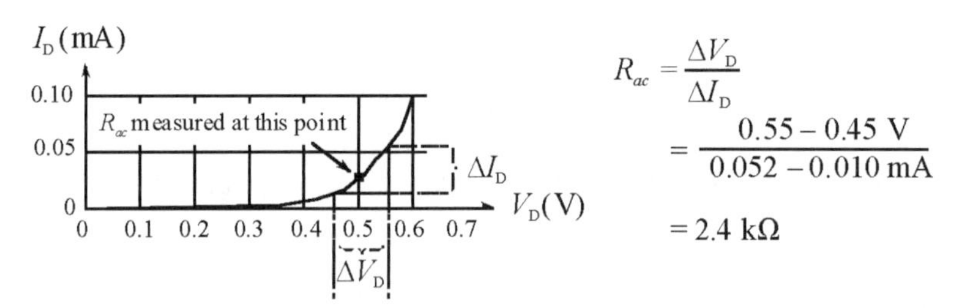

Figure 2-2 Determination of ac resistance.

In this experiment, you will take data on a forward- and a reverse-biased diode and plot the *I-V* characteristic. You will also set up a circuit to plot the diode response directly on an oscilloscope.

Materials Needed:

Resistors: one 330 Ω, one 1.0 M

One signal diode (1N914 or equivalent)

Procedure:

1. Measure and record the values of the resistors listed in Table 2-1. Then check your diode with an ohmmeter by measuring the forward- and reverse-resistance[1] by reversing the meter leads across the diode. The diode passes this test if the resistance is significantly different between the two measurements. If you are using an autoranging meter, the ohmmeter may not produce enough voltage to overcome the barrier potential; in this case, consult the operator's manual for specific instructions to down range the meter. Record the data in Table 2-1.

Table 2-1

Component	Listed Value	Measured Value
R_1	330 Ω	
R_2	1.0 MΩ	
D_1 forward resistance		
D_1 reverse resistance		

[1] Some DMMs have a special "diode test" position that sources a small current and displays the forward voltage drop in one direction and "OL" (overload) in the other direction for a good diode. Check your operator's manual if you have this provision.

2. Construct the forward-biased circuit shown in Figure 2-3. The line on the diode indicates the cathode side of the diode (with forward bias, this is the negative side). Set the power supply for zero volts.

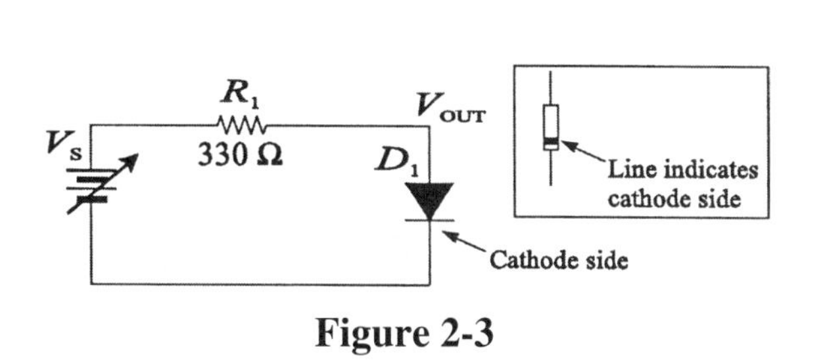

Figure 2-3

Table 2-2

V_F	V_{R1} (measured)	I_F (computed)
0.45 V		
0.50 V		
0.55 V		
0.60 V		
0.65 V		
0.70 V		
0.75 V		

3. Monitor the forward voltage drop, V_F, across the diode. Slowly increase V_S to establish 0.45 V across the diode. Measure the voltage across the resistor, V_{R1}, and record it in Table 2-2.

4. The diode forward current, I_F, can be found by applying Ohm's law to R_1. Compute I_F and enter the computed current in Table 2-2.

5. Repeat steps 3 and 4 for each voltage listed in Table 2-2.

6. With the power supply set to a voltage that causes 0.75 V to drop across the diode, bring a hot soldering iron near the diode. Do NOT touch the diode with the iron. Observe the effect of heat on the voltage and current in a forward-biased diode. If you have freeze spray available, test its effect on the diode's operation. Describe your observations.

____

7. The data in this step will be accurate only if your voltmeter has a high input impedance. You can find out if your meter is high impedance by measuring the power supply voltage through a series 1.0 MΩ resistor. If the meter reads the supply voltage accurately, it has high input impedance. Connect the reverse-biased circuit shown in Figure 2-4. Set the power supply to each voltage listed in Table 2-3. Apply Ohm's law to the resistor and compute the reverse current in each case. Enter the computed current in Table 2-3.

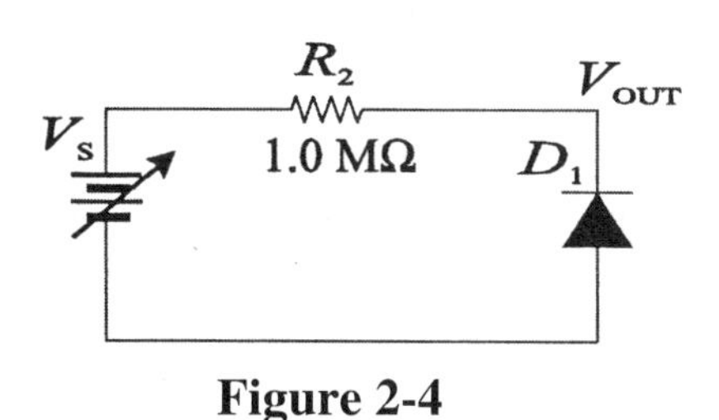

Figure 2-4

Table 2-3

V_{R2} (measured)	V_R (measured)	I_R (computed)
5.0 V		
10.0 V		
15.0 V		

8. Graph the forward- and reverse-biased diode curves on Plot 2-1. The different voltage scale factors for the forward and reverse curves are chosen to allow the data to cover more of the graph. You need to choose an appropriate current scale factor which will put the largest current recorded near the top of the graph. Suggestions for graphing are given on page 2 of the Introduction to the Student. The first two steps have been completed for the plot shown.

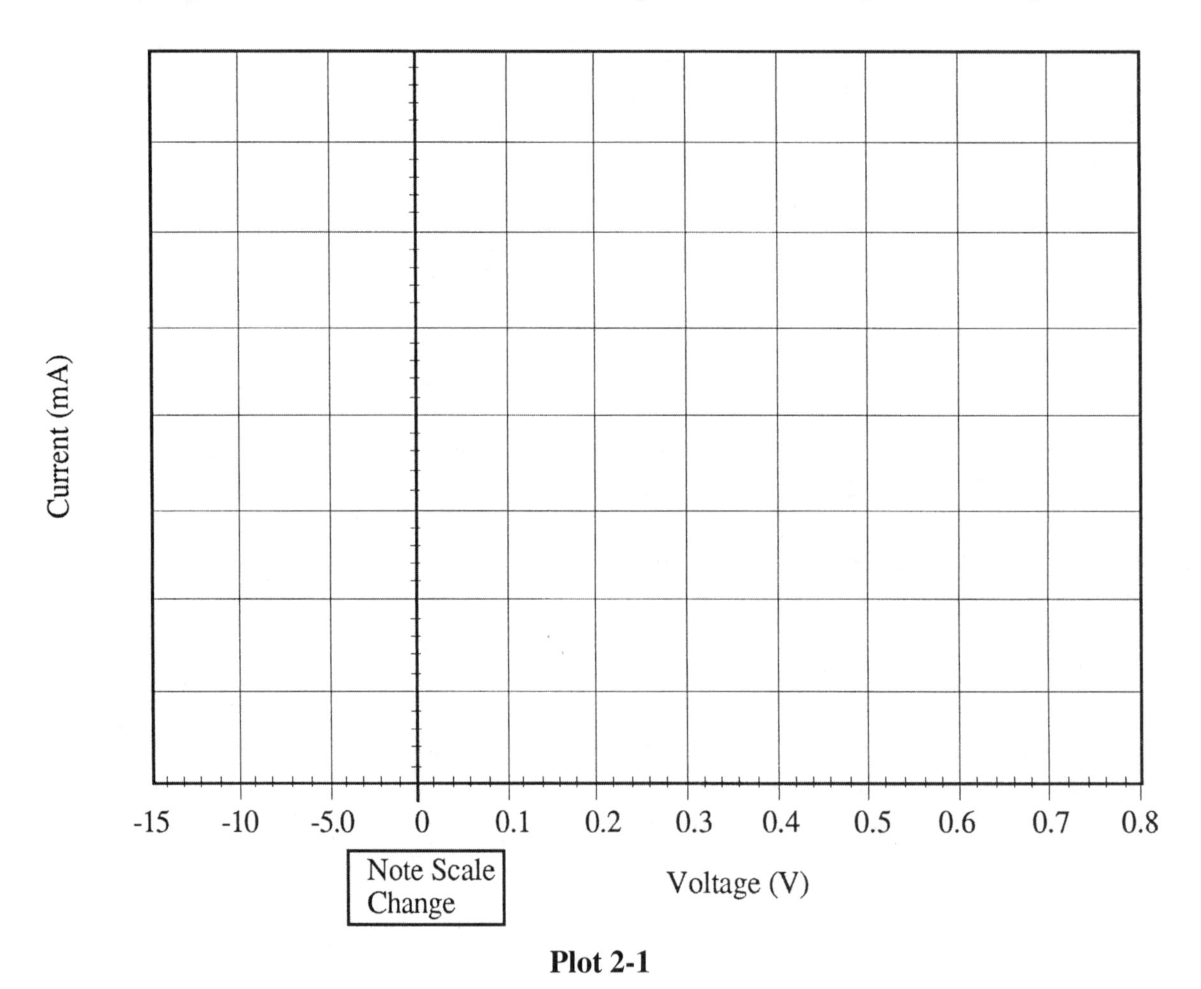

Plot 2-1

9. With the power supply set to 15 V, bring a hot soldering iron near the diode. Don't touch the diode with the iron. Observe the effect of heat on the reverse-biased diode. If you have freeze spray available, test its effect on the diode's operation. Describe your observations.

____

____

10. You can plot the diode's forward characteristic on your oscilloscope by connecting the circuit shown in Figure 2-5. Channel 1 senses the voltage drop across the diode; channel 2 shows a signal that is proportional to the current. The scope is placed in the X-Y mode. The signal generator ground must *not* be the same as the scope ground. Channel 2 must be inverted to display the signal in the proper orientation. Set up the circuit and observe the signal. Discuss your results in the conclusion statement for this experiment.

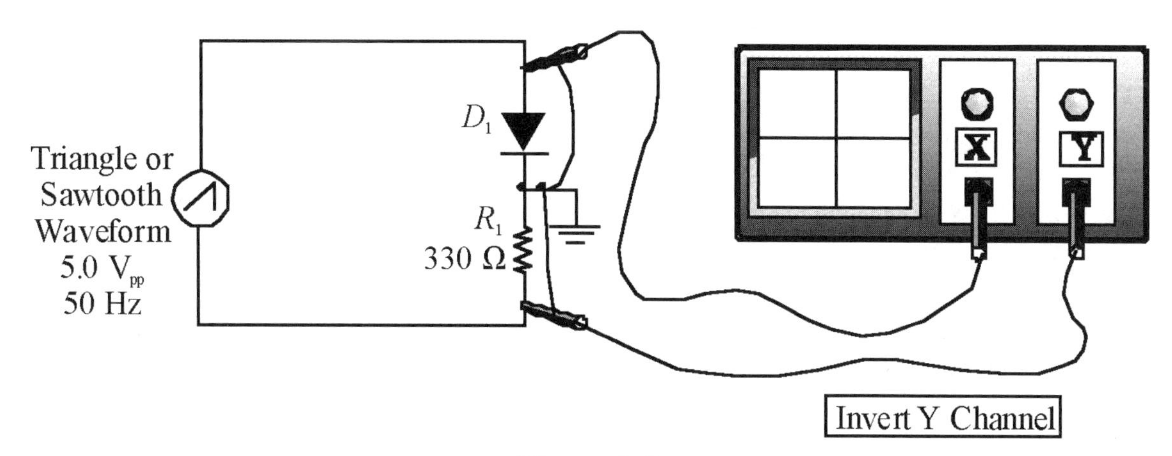

Figure 2-5

Conclusion:

Evaluation and Review Questions:

1. What factors affected the accuracy of the measurements in this experiment? (Consider both the forward-biased and reverse-biased cases.)

2. Compute the diode's ac resistance at three points on the forward-biased curve. Apply Ohm's law to the curve in Plot 2-1 at 0.5 V, 0.6 V, and 0.7 V by dividing a small change in voltage by a small change in current, as illustrated in Figure 2-2.

R_{ac} (0.5 V) = _______ R_{ac} (0.6 V) = _______ R_{ac} (0.7 V) = _______

3. From the data in Table 2-2, compute the maximum power dissipated in the diode.

4. Based on your observations of the heating and cooling of a diode, what does heat do to the forward and reverse resistance of a diode?

5. Explain how you could use an ohmmeter to identify the cathode of an unmarked diode. Why is it necessary to know the actual polarity of the ohmmeter leads?

For Further Investigation:

The theoretical equation for a diode's *I-V* curve shows that the current is an exponential function of the bias voltage. This means that the theoretical forward diode curve will plot a straight-line on semilog paper. Semilog paper contains a logarithmic scale on one axis and a linear scale on the other axis. Add the proper labels and graph your data from this experiment (Table 2-2) onto Plot 2-2. What conclusion can you make from the data you recorded?

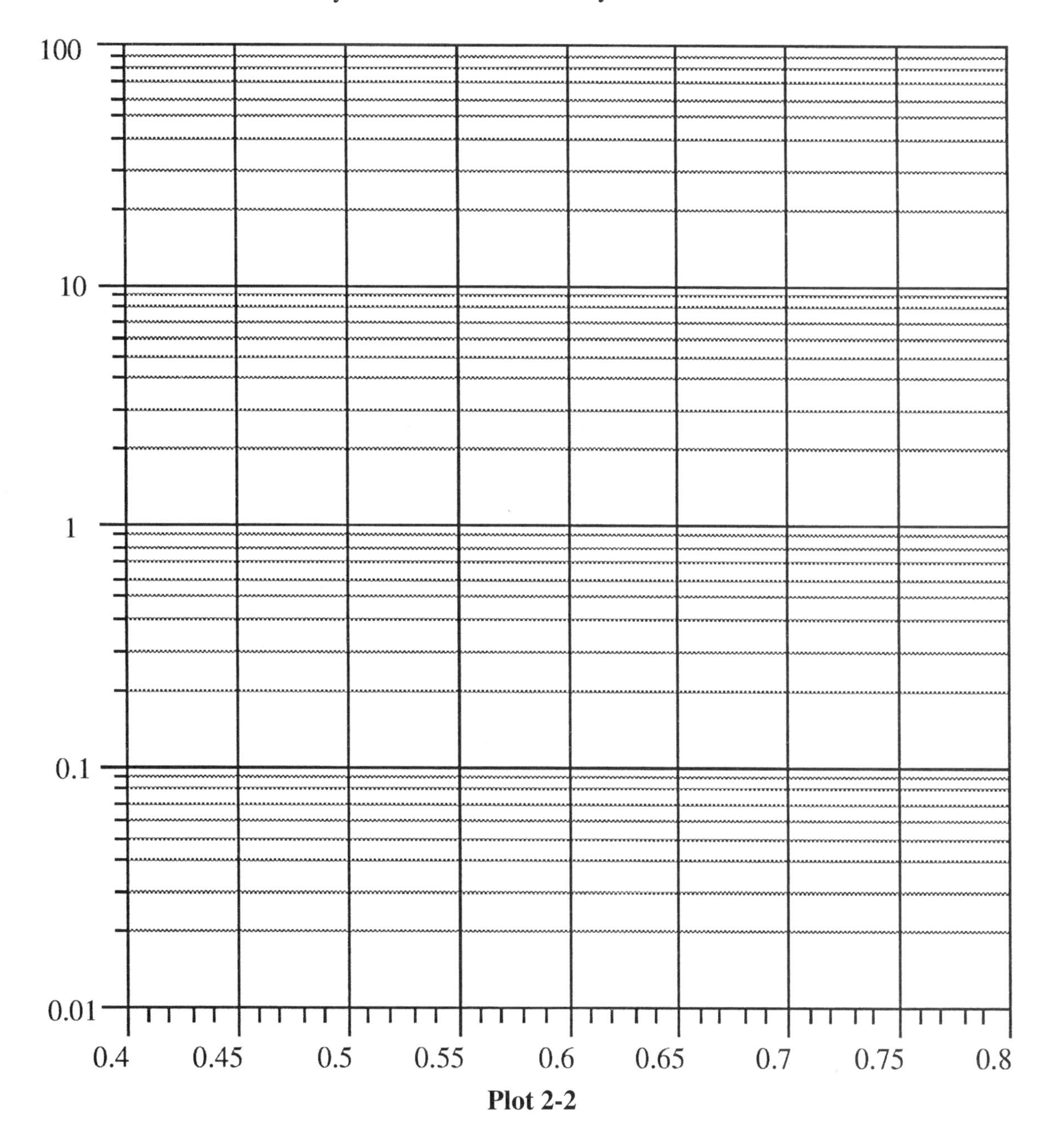

Plot 2-2

Name ______________
Date ______________
Class ______________

3 Rectifier Circuits

Reading:
Floyd and Buchla, *Analog Fundamentals: A Systems Approach*, Sections 2-5 through 2-6

Objectives:
After performing this experiment, you will be able to:
1. Construct half-wave, full-wave, and bridge rectifier circuits, and compare the input and output voltage for each.
2. Connect a filter capacitor to each circuit in objective 1 and measure the ripple voltage and ripple frequency.

Summary of Theory:
Rectifiers are diodes used to change ac into dc. As you saw in Experiment 1, diodes work like a one-way valve, allowing current in only one direction. When ac is applied to a diode, the diode is forward-biased for one-half of the cycle and reverse-biased for the other half cycle. The output waveform is a pulsating dc waveform (or *half-wave rectified*) as illustrated in Figure 3-1. This pulsating dc waveform can then be filtered to convert it to constant dc.

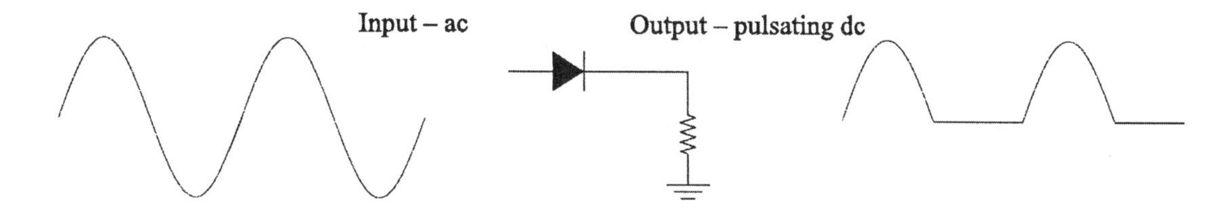

Figure 3-1

Rectifiers are widely used in power supplies to provide the dc voltage necessary for almost all active devices to work. The three basic rectifier circuits are the half-wave, the center-tapped full-wave, and the full-wave bridge rectifier circuits. The most important parameters for choosing diodes for these circuits are the maximum forward current, I_F, and the peak inverse voltage rating (PIV) of the diode. The peak inverse voltage is the maximum voltage the diode can withstand when it is reverse-biased. The amount of reverse voltage that appears across a diode depends on the type of circuit in which it is connected. Some characteristics of the three rectifier circuits will be investigated in this experiment.

Materials Needed:
Resistors: two 2.2 kΩ resistors
One 12.6 V ac center-tapped transformer with fused line cord
Four diodes 1N4001 (or equivalent)
One 100 μF capacitor
For Further Investigation:
One 0.01 μF capacitor
One 7812 or 78L12 regulator

Procedure:

1. Connect the half-wave rectifier circuit shown in Figure 3-2. (**Safety note** – the ac line voltage must not be exposed; the transformer should be fused as shown.) Notice the polarity of the diode. The line indicates the cathode side (the negative side when forward-biased). Connect the oscilloscope so that channel 1 is across the transformer secondary and channel 2 is across the output (load) resistor. The oscilloscope should be set for LINE triggering as the waveforms to be viewed in this experiment are synchronized with the ac line voltage. View the input voltage, V_{sec}, and output voltage, V_{out}, waveforms for this circuit and sketch them on Plot 3-1 below. Label voltage and time on your sketch.

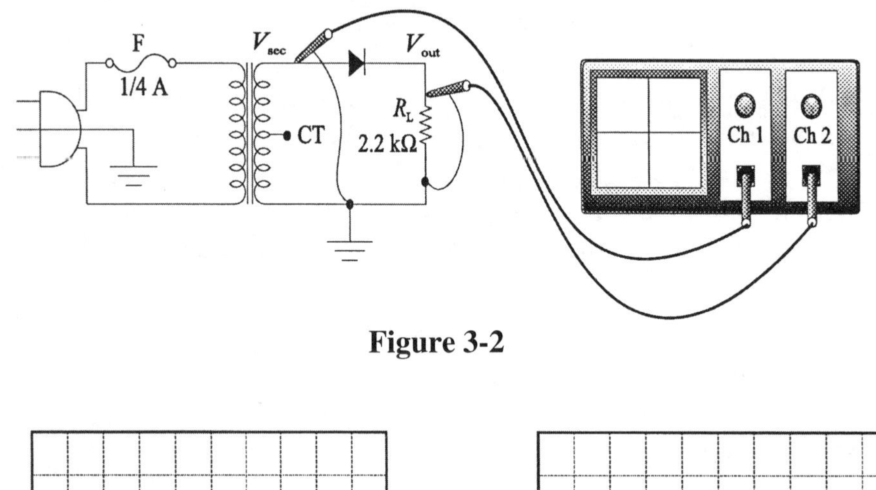

Figure 3-2

V_{sec}

V_{out}
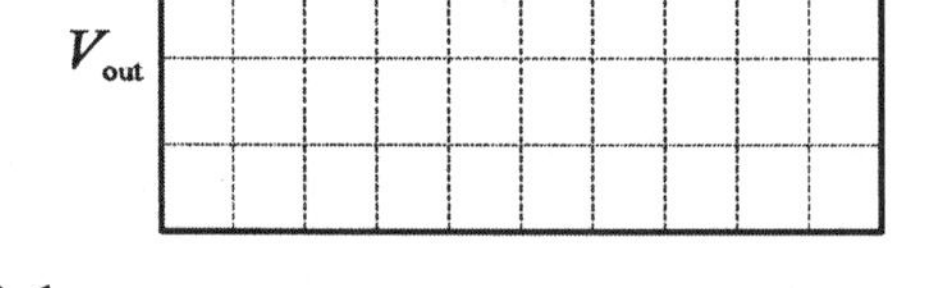

Plot 3-1

2. Measure the secondary rms voltage and the output peak voltage. Remember to convert the oscilloscope reading to rms voltage. Record the data in Table 3-1.

3. The output isn't very useful as a dc voltage source because of the pulsating output. Connect a 100 μF filter capacitor in parallel with the load resistor (R_L). Check the polarity of the capacitor; the negative side goes toward ground. Measure the dc load voltage, $V_{out(DC)}$, and the peak-to-peak ripple voltage, $V_{r(pp)}$, in the output. To measure the ripple voltage, switch the oscilloscope to AC COUPLING. This allows you to magnify the small ac ripple voltage without

including the much larger dc level. Measure the ripple frequency. The ripple frequency is the frequency at which the waveform repeats. Record all data in Table 3-1.

Table 3-1 Half-Wave Rectifier

Without Filter Capacitor				With Filter Capacitor		
Computed	Measured	Computed	Measured	Measured		
$V_{sec(rms)}$	$V_{sec(rms)}$	$V_{out(p)}$	$V_{out(p)}$	$V_{out(DC)}$	$V_{r(pp)}$	Ripple Frequency
12.6 V ac						

4. Disconnect power and change the circuit to the full-wave rectifier circuit shown in Figure 3-3. Notice that the ground for the circuit has changed. The oscilloscope ground needs to be connected as shown. Check your circuit carefully before applying power. Compute the expected peak output voltage. Then apply power and view the V_{sec} and V_{out} waveforms. Sketch the observed waveforms on Plot 3-2.

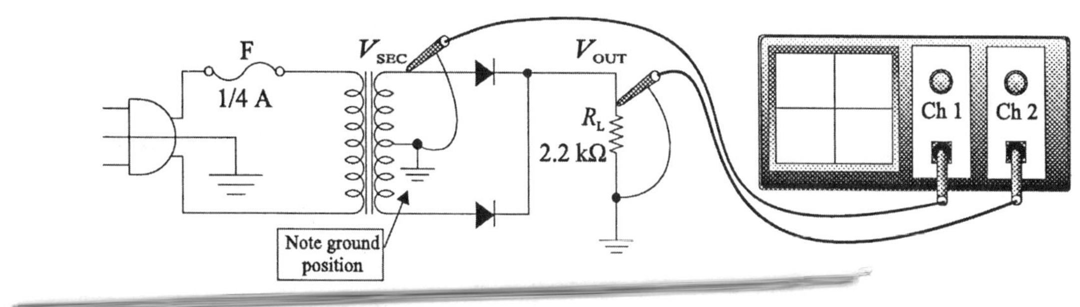

Figure 3-3

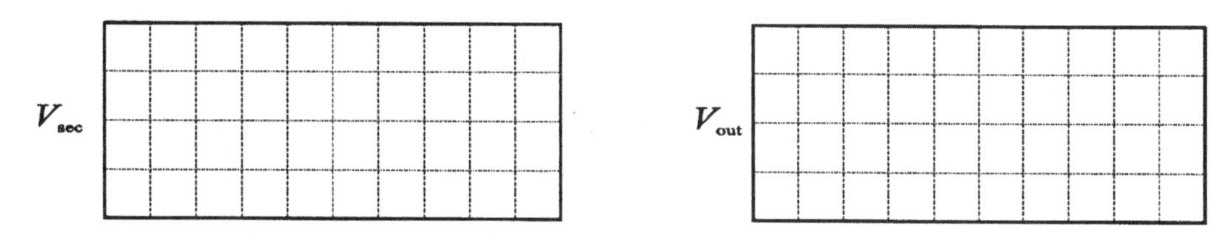

Plot 3-2

5. Measure $V_{sec(rms)}$ and the peak output voltage ($V_{out(p)}$) without a filter capacitor. Record the data in Table 3-2.

6. Now add a 100 μF capacitor in parallel with the load resistor. Measure $V_{out(DC)}$, the peak-to-peak ripple voltage, $V_{r(pp)}$, and the ripple frequency as before. Record the data in Table 3-2.

Table 3-2 Full-Wave Rectifier Circuit

Without Filter Capacitor				With Filter Capacitor		
Computed	Measured	Computed	Measured	Measured		
$V_{sec(rms)}$	$V_{sec(rms)}$	$V_{out(p)}$	$V_{out(p)}$	$V_{out(DC)}$	$V_{r(pp)}$	Ripple Frequency
6.3 V ac						

7. Investigate the effect of the load resistor on the ripple voltage by connecting a second 2.2 kΩ load resistor in parallel with R_L in the full-wave circuit in Figure 3-3. The filter capacitor is not shown but should be left in parallel also. Measure the ripple voltage. What can you conclude about the effect of additional load current on the ripple voltage?

____

8. Disconnect power and change the circuit to the bridge rectifier circuit shown in Figure 3-4. Notice that no terminal of the transformer secondary is at ground potential. The input voltage to the bridge, V_{sec}, is not referenced to ground. *The oscilloscope cannot be used to view both the secondary voltage and the output voltage at the same time.* Check your circuit carefully before applying power. Compute the expected peak output voltage. Then apply power and *use a voltmeter* to measure $V_{sec(rms)}$. Use the oscilloscope to measure the peak output voltage ($V_{out(p)}$) without a filter capacitor. Record the data in Table 3-3.

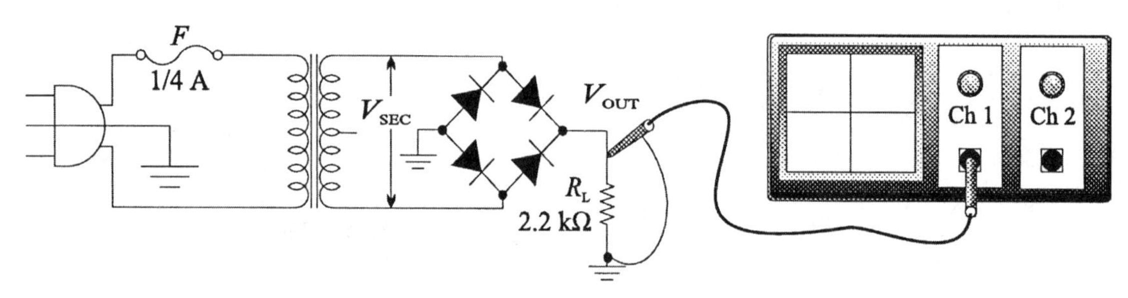

Figure 3-4

Table 3-3 Bridge Rectifier Circuit

Without Filter Capacitor				With Filter Capacitor		
Computed	Measured	Computed	Measured	Measured		
$V_{sec(rms)}$	$V_{sec(rms)}$	$V_{out(p)}$	$V_{out(p)}$	$V_{out(DC)}$	$V_{r(pp)}$	Ripple Frequency
12.6 V ac						

9. Connect the 100 μF capacitor in parallel with the load resistor. Measure $V_{out(DC)}$, the peak-to-peak ripple voltage, and the ripple frequency as before. Record the data in Table 3-3.

10. Simulate an open diode in the bridge by removing one diode from the circuit. What happens to the output voltage? The ripple voltage? The ripple frequency?

____

Conclusion:

Evaluation and Review Questions:

1. What advantage does a full-wave rectifier circuit have over a half-wave rectifier circuit?

2. Compare a bridge rectifier circuit with a full-wave rectifier circuit. Which has the higher output voltage? Which has the greater current in the diodes?

3. In step 4, you moved the ground reference to the center-tap of the transformer. If you wanted to look at the voltage across the entire secondary, you would need to connect the oscilloscope as shown in Figure 3-5 and *add* the two channels. (Some oscilloscopes do not have this capability.) Why is it necessary to use *two* channels to view the entire secondary voltage?

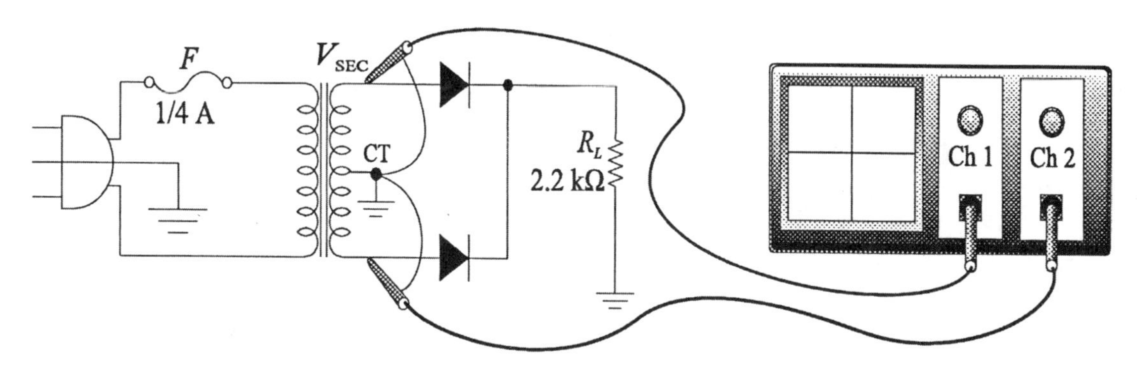

Figure 3-5

4. Explain how you could measure the ripple frequency to determine if a diode were open in a bridge rectifier circuit.

5. (a) What is the maximum dc voltage you could expect to obtain from a transformer with an 18 Vrms secondary using a bridge rectifier circuit with a filter capacitor?

   (b) What is the maximum dc voltage you could expect to obtain from the same transformer connected in a full-wave rectifier circuit with a filter capacitor?

For Further Investigation:
The bridge rectifier circuit shown in Figure 3-4 can be readily changed to a +12 V regulated power supply with the addition of a 7812 or 78L12 three-terminal regulator. These three-terminal regulators are easy to use and provide a regulated +12 V output. The 7812 can deliver over 1.0 A of current while the 78L12 can deliver over 100 mA. Add one of the regulators to your bridge rectifier circuit as shown in Figure 3-6. The 100 μF filter capacitor should be connected across the input of the regulator and a 0.01 μF capacitor placed across the output as shown. Measure the output ripple from the circuit with the regulator. Compare your results with the unregulated circuit in step 9.

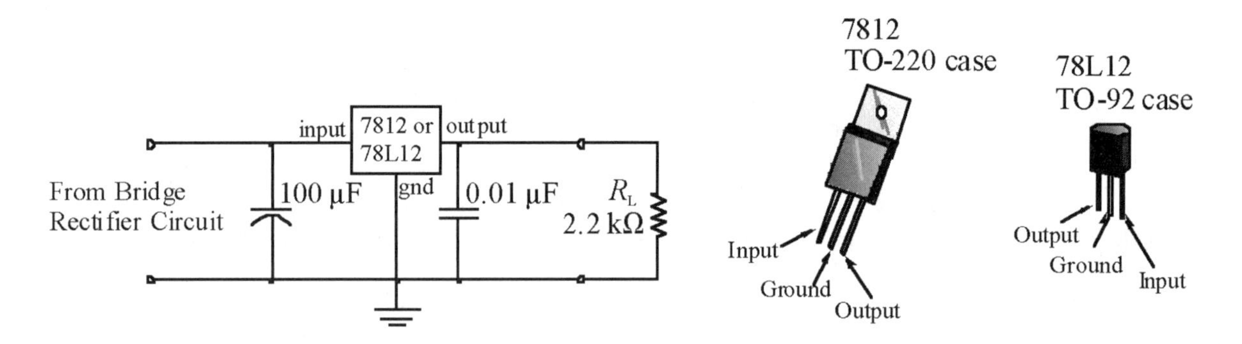

Figure 3-6

	Name ______________
	Date ______________
	Class ______________

4 Diode Limiting and Clamping Circuits

Reading:
Floyd and Buchla, *Analog Fundamentals: A Systems Approach*, Section 2-7

Objectives:
After performing this experiment, you will be able to:
1. Explain the difference between limiting and clamping circuits.
2. Calculate and measure the voltage limits of both biased and unbiased limiting circuits.
3. Predict and measure the effect of a dc bias voltage on a clamping circuit.

Summary of Theory:
Diodes are frequently used in applications such as waveshaping, mixers, detectors, protection circuits, and switching circuits. In this experiment, you will investigate two widely used applications of diode circuits, diode *limiting* circuits and diode *clamping* circuits. Diode limiting circuits (also called *clipping circuits*) are used to prevent a waveform from exceeding some particular limits, either negative or positive. For example, assume it is desired to remove the portion of sine wave that exceeds +5.0 V. The bias voltage, V_{BIAS}, is set to a voltage 0.7 V less than the desired clipping level. The circuit in Figure 4-1 will limit the waveform because the diode will be forward-biased whenever the signal exceeds +5.0 V. This places V_{BIAS} in parallel with R_L and prevents the input voltage from going above +5.0 V. When the signal is less than +5.0 V, the diode is reverse-biased and appears to be an open circuit. If, instead, it was desired to clip the waveform below some specified level, the diode can be reversed and V_{BIAS} is set to 0.7 V greater than the desired limiting level.

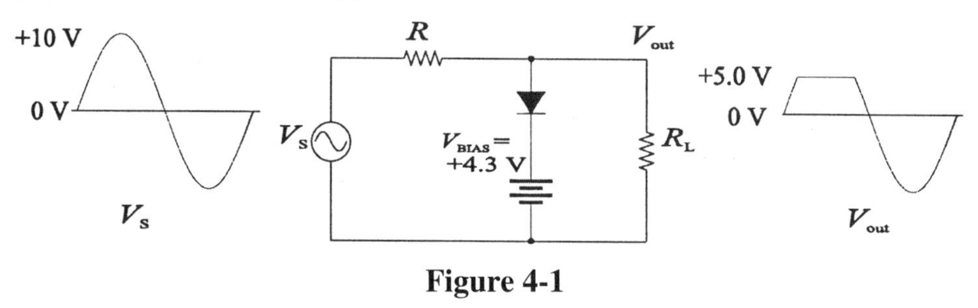

Figure 4-1

Diode clamping circuits are used to shift the dc level of a waveform. If a signal has passed through a capacitor, the dc component is blocked. A clamping circuit can restore the dc level. For this reason these circuits are sometimes called *dc restorers.* Diode clamping action is illustrated in Figure 4-2 for both positive and negative clamping circuits. The diode causes the series capacitor to have a low-resistance charging path and a high-resistance discharge path through R_L. As long as the *RC* time

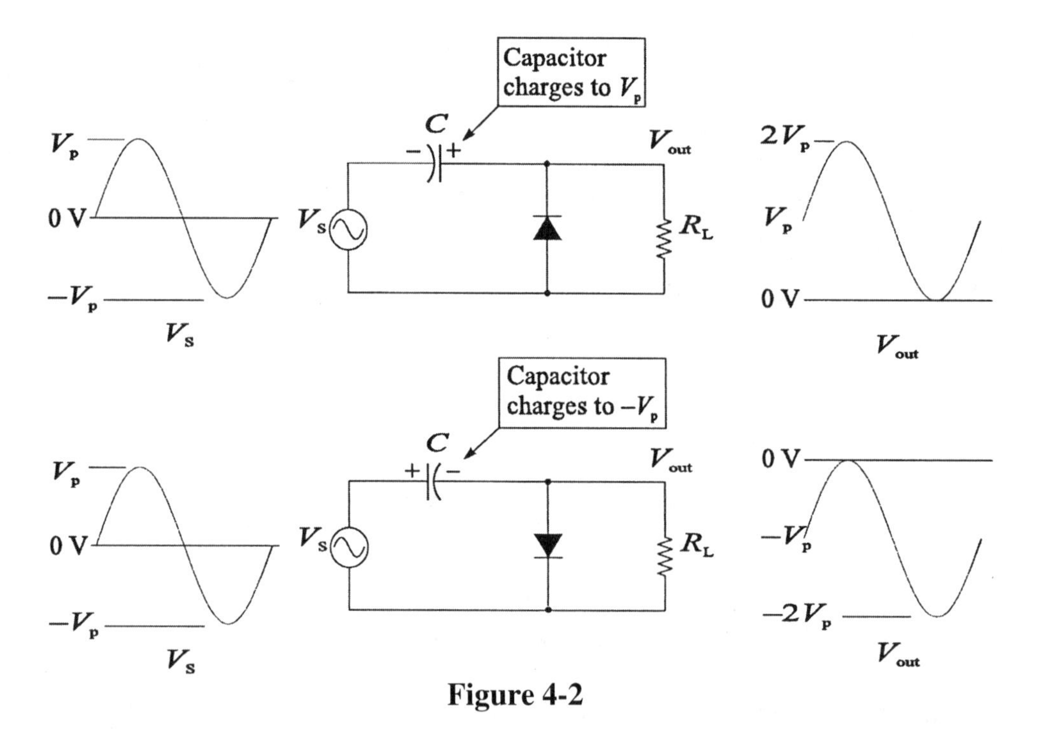

Figure 4-2

constant is long compared to the period of the waveform, the capacitor will be charged to the peak value of the input waveform. This action requires several cycles of the input signal to charge the capacitor. The output load resistor sees the sum of the dc level on the capacitor and the input voltage.

Materials Needed:

Resistors: two 10 kΩ, one 47 kΩ
Two signal diodes: 1N914 (or equivalent)
One 47 μF capacitor
For Further Investigation:
Three 1.0 kΩ resistors

Procedure:

1. Connect the circuit shown in Figure 4-3. Connect the signal generator to the circuit and set it for a 6.0 V_{pp} sine wave at a frequency of 1.0 kHz with no dc offset. Observe the input and output waveforms on the oscilloscope by connecting it as shown. Notice that R_2 and R_L form a voltage divider, causing the load voltage to be less than the source voltage. R_1 will provide a dc return path in case the signal generator is capacitively coupled.

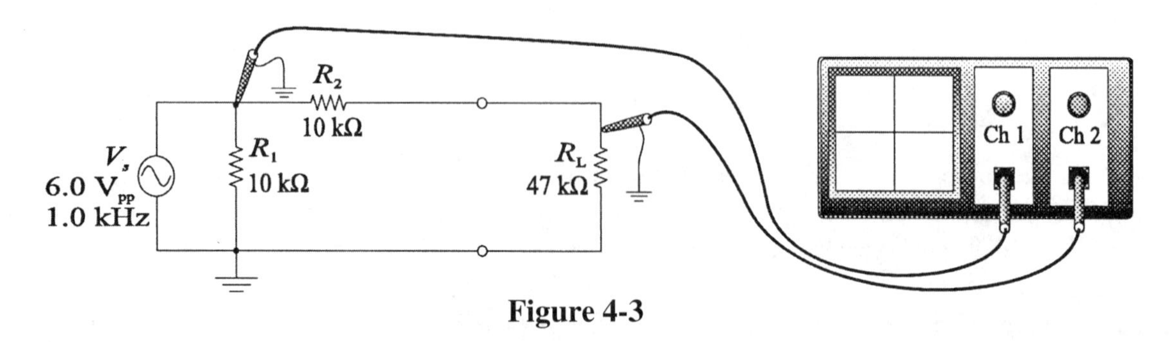

Figure 4-3

2. Add the diode to the circuit as shown in Figure 4-4. Look carefully at the output waveform. Notice the zero volt level. Sketch the input and output waveforms in the space provided. Then measure[1] the waveform across R_2. Sketch the observed waveforms on Plot 4-1.

Figure 4-4

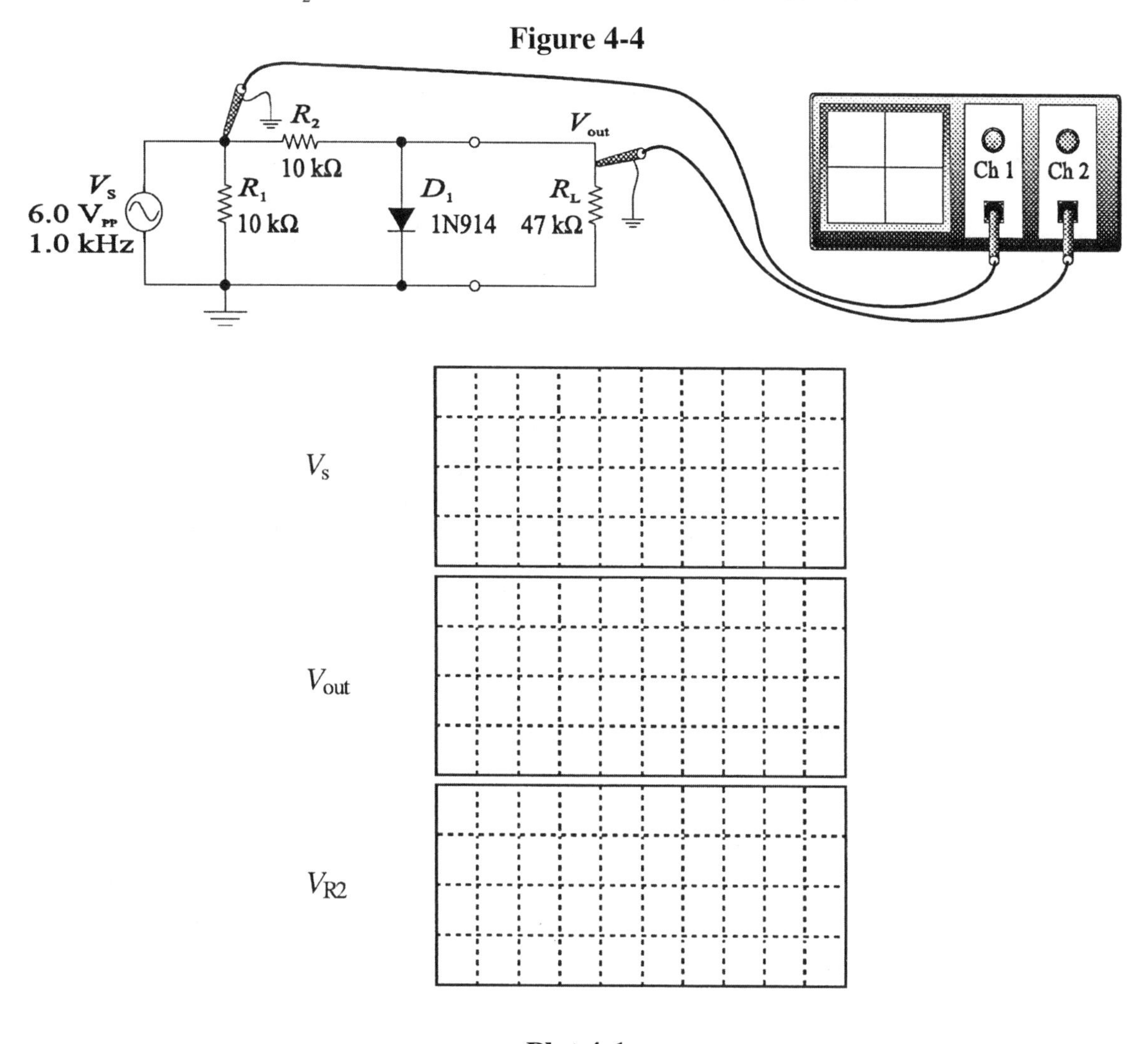

Plot 4-1

3. Remove the cathode of the diode from ground and connect it to the power supply as shown in Figure 4-5. Vary the voltage from the supply and describe the results.

____

[1] On the oscilloscope, this is accomplished by placing the probes from each channel on both sides of R_2. The channels are set to the same vertical sensitivity (VOLTS/DIV). For most scopes, you then select the ADD function and INVERT CH-2. Check with your operator's manual if these functions are not available.

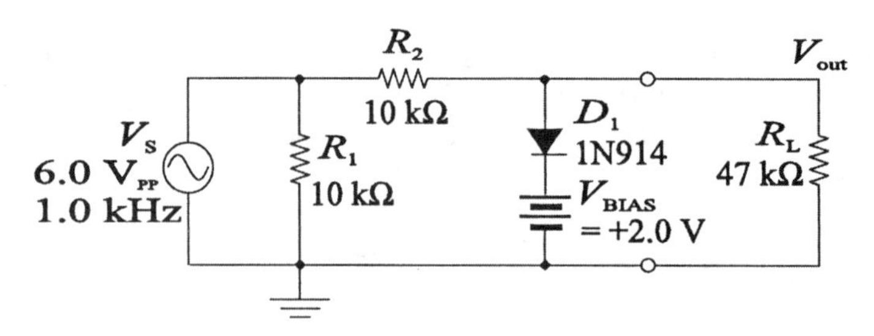

Figure 4-5

4. Reverse the diode in the circuit of Figure 4-5. Vary the dc voltage and describe the results.

____

5. Replace the positive power supply with a negative supply. Again, vary the dc voltage and describe the results.

____

6. If you have freeze spray available, test the effect on the clipping level when the diode is cooled. Observations:

____

7. Connect the clamp circuit shown in Figure 4-6. Couple the oscilloscope with dc coupling and observe the output voltage. Vary the input voltage. Observations:

____

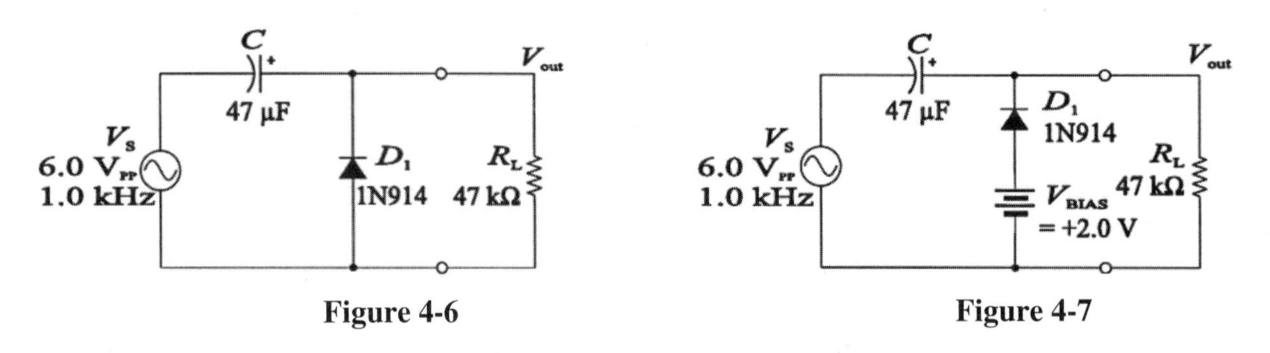

Figure 4-6 **Figure 4-7**

8. Add a dc voltage to the diode by connecting the power supply as shown in Figure 4-7. Sketch the output waveform on Plot 4-2 below. Show the dc level on your sketch:

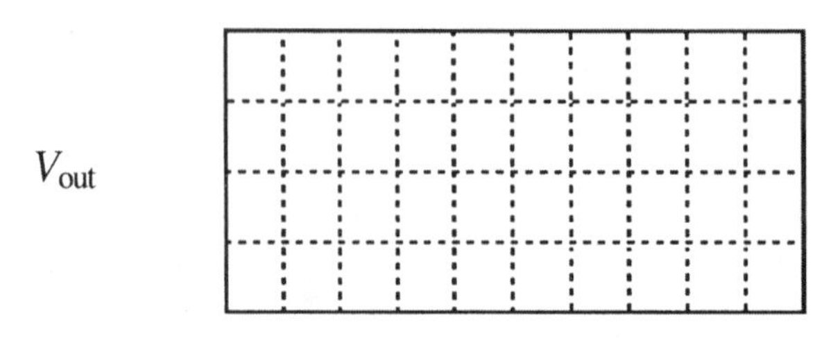

Plot 4-2

9. Find out what happens if the positive dc voltage is replaced with a negative dc source.

Observations:____

Conclusion:

Evaluation and Review Questions:

1. In step 2, you observed the voltage waveform across the series resistor, R_2. The waveform observed across R_2 could have been predicted by applying Kirchhoff's voltage law to V_s and V_{out}. Explain.

2. For the circuit of Figure 4-6, describe what would happen to the output voltage if the capacitor were shorted.

3. For the circuit of Figure 4-7, what change would you expect in the output if the diode were reversed?

4. Explain the difference between a limiting and a clamping circuit.

5. Sketch the output waveform for the limiting circuit shown in Figure 4-8.

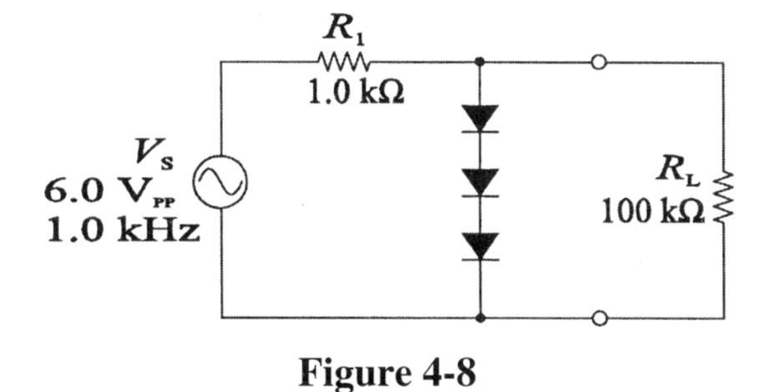

Figure 4-8

For Further Investigation:

Suppose you wanted to set up a limiting circuit that would clip signals above +5.0 V but had only a fixed +10 V supply available. You might try setting up the voltage divider shown in Figure 4-9 to set the limiting level. Build the circuit and drive it with a 20 V_{pp} sinusoidal wave. Can you explain why the output is not flat as in Figure 4-1? Try putting the 47 μF capacitor across R_2. What happens? Why?

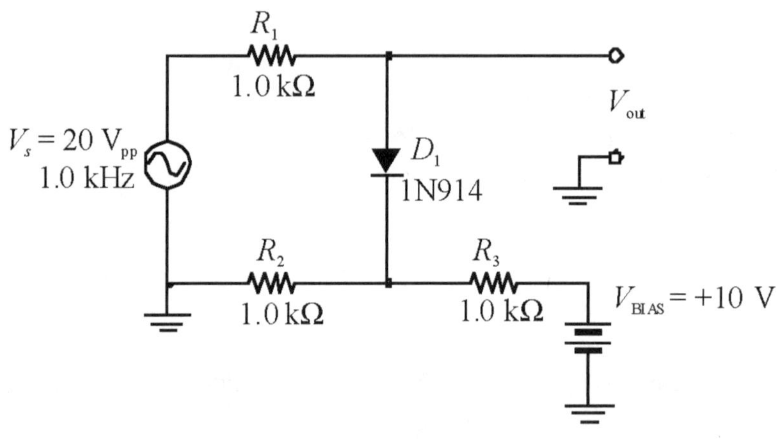

Figure 4-9

Name ______________
Date ______________
Class ______________

5 Special-Purpose Diodes

Reading:
Floyd and Buchla, *Analog Fundamentals: A Systems Approach*, Section 2-8

Objectives:
After performing this experiment, you will be able to:

1. Measure the resonant frequency of a voltage-controlled resonant circuit containing a varactor diode as a function of the bias voltage. Plot the response of the circuit.
2. Measure the Q of the resonant circuit in objective 1.
3. Measure the I-V characteristic for three light-emitting diodes (LEDs) and for a photocell.

Summary of Theory:
A number of diodes are designed for special applications. In this experiment, you will investigate three special-purpose diodes: the varactor diode, the light-emitting diode (LED), and the photodiode. A varactor is a reverse-biased diode that can be used as a small capacitor in applications such as tuners. The *n*- and *p*-layers of the diode are conductive regions, similar in this application to the conductive plates of a capacitor. The depletion region acts as the dielectric. The capacitance varies inversely with the width of the depletion region, which is controlled by the reverse-bias. Although a forward-biased diode has capacitance also, it would not act as a capacitor because it does not block dc. In the first part of the experiment, you will construct a resonant circuit using a varactor and measure the resonant frequency as the bias voltage is changed.

An LED is a special-purpose diode that emits infrared (invisible to the eye) or visible light depending on the type. Like other diodes, LEDs allow current in only one direction. They are used in a wide variety of displays and as small, highly reliable light sources for optical sensors that respond to an object breaking or reflecting a beam of light. LEDs are rated in terms of electrical and optical characteristics. Electrical characteristics include the forward voltage drop at a specified forward current, reverse leakage current I_R, and reverse breakdown voltage $V_{(BR)R}$. Optical characteristics include the peak emission wavelength (in nm) and power output, P_O, (in milliwatts or microwatts) and spectral output. Depending on the construction, LEDs can be made to radiate from the near-ultraviolet to the near infrared in a narrow band of frequencies. LEDs are not constructed from silicon or germanium like ordinary diodes, but are made from compounds of gallium arsenic and phosphorous. As such, they do not have the same forward-bias characteristic of silicon devices. You will measure the I-V characteristic and compare different colored LEDs.

Photodiodes are special diodes that are used as photosensors. When they are reverse-biased, the depletion region is increased. Light that strikes the depletion region causes hole-electron pairs to be formed, causing current in the external circuit. The amount of current is dependent on the wavelength and power density of the light as well as the sensitivity of the diode.

A *phototransistor* is a photodiode coupled to an internal transistor. This increases the sensitivity to light but has a slower response than a photodiode alone. In the For Further Investigation, you will use a phototransistor as a sensor to test the radiation pattern of an LED. Later, in Experiment 38, phototransistors will be tested again as part of a motion sensor.

Materials Needed:

Resistors: one 1.0 kΩ, one 1.0 MΩ
One 10 kΩ potentiometer
Two MV2115 varactors
One MRD500 photodiode
One 15 mH inductor
One small transformer with a 12.6 V secondary
Three LEDs, one red, one yellow, one green
Light source (bright lamp or flashlight)

For Further Investigation:

One MRD300 phototransistor
Resistors: one 510 Ω, one 330 kΩ
Masking tape
Heat shrink tubing (for light baffle) that fits over phototransistor

Procedure:

Part 1: Varactor diode

1. Construct the varactor resonant circuit shown in Figure 5-1. The circuit uses two reverse-biased varactors to prevent either from becoming forward-biased; however, at zero volts of bias, the lower varactor is shorted by the potentiometer. Adjust the bias voltage to zero volts. Using the oscilloscope, set the function generator to 1.0 V_{pp} at a frequency of 50 kHz. Observe the signal across the inductor. At this frequency, the amplitude will be quite small. Increase the frequency slowly, and you should observe the signal rise suddenly to a maximum, then fall. The frequency of the maximum signal is the resonant frequency. Measure the resonant frequency and record it in the first space in Table 5-1.

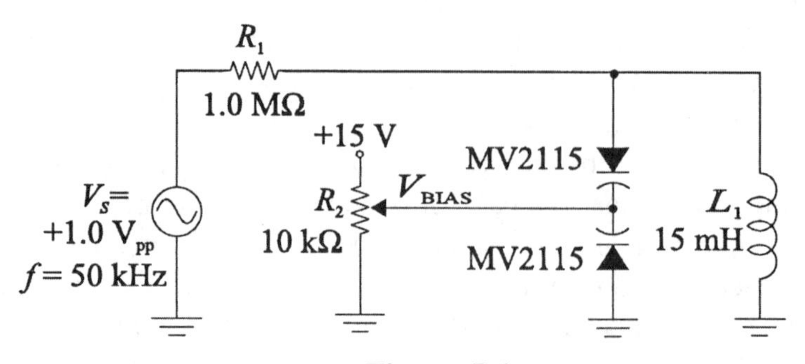

Figure 5-1

Table 5-1

V_{BIAS}	Resonant Frequency, f_r
0.0 V	
1.0 V	
2.0 V	
4.0 V	
8.0 V	
15.0 V	

2. Increase the bias voltage to +1.0 V. Measure and record the new resonant frequency. Continue in this manner for each bias voltage given in Table 5-1.

3. Plot the response of the circuit in Plot 5-1. The independent variable is bias voltage so it should be plotted on the *x*-axis, and resonant frequency should be plotted on the *y*-axis. Add the proper labels to your graph.

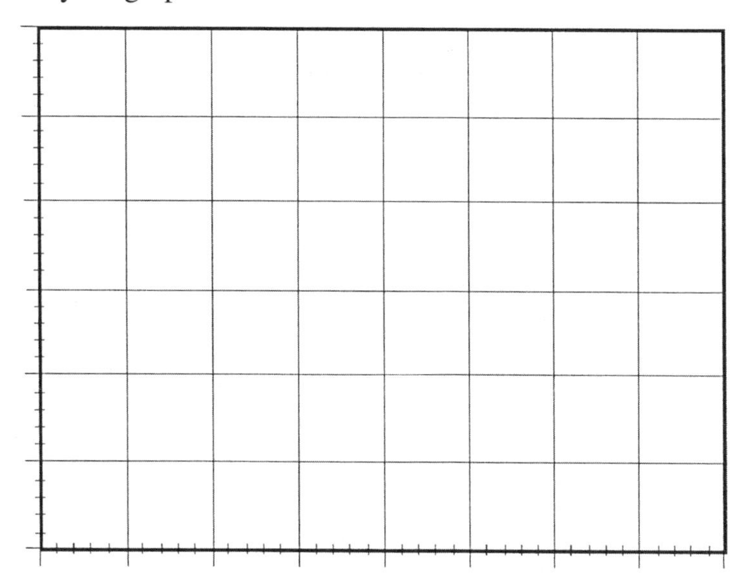

Plot 5-1

4. Measure the Q of the varactor circuit by testing the response of the circuit. Leave the bias voltage at a constant +15 V. Record the resonant frequency, f_r, you found previously for +15 V of bias in Table 5-2. Then raise the frequency of the generator until the voltage across the inductor drops to 70.7% of the maximum (at resonance). This is the upper critical frequency, f_{cu}. Measure and record this frequency in Table 5-2.

Table 5-2

Parameter	Measured Value
resonant frequency, f_r	
upper critical frequency, f_{cu}	
lower critical frequency, f_{cl}	
bandwidth, BW	
Q	

5. Reduce the frequency of the generator until the voltage across the inductor drops to 70.7% of the maximum. This is the lower critical frequency, f_{cl}. Measure and record this frequency in Table 5-2. Then compute the bandwidth, BW, for the circuit by subtracting f_{cl} from f_{cu}. Find Q for the circuit from the relation:

$$Q = \frac{f_r}{BW}$$

Part 2: Light-emitting diode and photodiode

1. Test the *I-V* characteristic of three different colored LEDs: red, yellow, and green. Do this by setting up the circuit shown in Figure 5-2. Note that the transformer secondary is *not* grounded. With the power off to the circuit, place the oscilloscope in the X-Y mode with Channel 2 inverted. When a dot is on the screen, you need to keep the intensity *very low* to avoid damage to the CRT. Position the dot at center screen. Set the X channel to 1.0 V/DIV and set the Y channel to 5.0 V/DIV. The X channel represents voltage across the diode and the Y channel represents current. Observe the *I-V* characteristic for each of the three LEDs and describe your observations. Note the variation in the forward-bias voltage and the slope of the forward-bias characteristic.

   red LED: ____

   yellow LED: ____

   green LED: ____

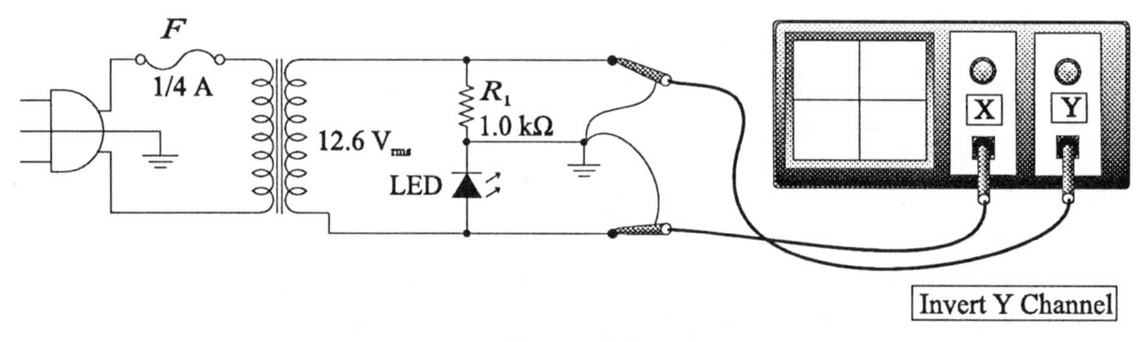

Figure 5-2

2. You can observe the characteristics of a photodiode with a circuit similar to the one you built in step 1. Since a typical photodiode sources current in the microamp region, change the resistor in Figure 5-2 to 1.0 MΩ and replace the LED with a MRD500 photodiode. (A larger resistor will give a higher voltage). Observe the reverse part of the characteristic curve as you shine a bright light into the photodiode. The photodiode is directional, so the light needs to be directly above the photodiode.

   Observations: ____

   ____

Conclusion:

Evaluation and Review Questions:

1. (a) At zero bias, only the upper varactor is active in the resonant circuit of Figure 5-1. Compute the capacitance of this diode at this point.

   (b) Based on the resonant frequency you observed for the varactor diode with +15 V of bias, compute the capacitance of *one* of the varactor diodes when the bias voltage was maximum. (Remember, when two capacitors are in series, the total capacitance is *less* than the capacitance of either one.)

2. Assume that resistor R_1 in Figure 5-1 was changed to 100 kΩ.
   (a) Would this have an effect on the resonant frequency? Explain.

   (b) Would this have an effect on Q? Explain.

3. Consider the circuit shown in Figure 5-3. Assume the LED drops 2.0 V when it is forward-biased.
   (a) Compute the current in the LED.

   (b) Assume the maximum permissible current in the LED is 30 mA. What value of R will produce this current?

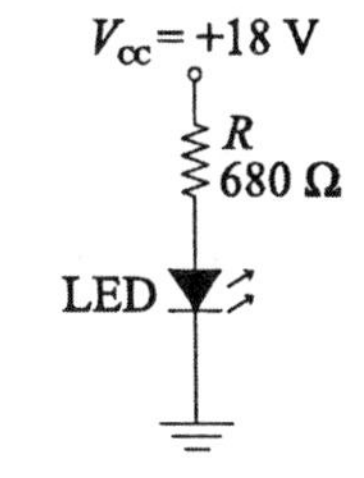

Figure 5-3

4. LEDs are rated in terms of electrical and optical characteristics. Give examples of each characteristic that could be found on a specification sheet.

5. The photodiode characteristic was observed using a 1.0 MΩ resistor instead of the 1.0 kΩ resistor that was used for the LED. Explain why this change was necessary.

For Further Investigation:

You can test the directional characteristics of an LED by making a detector using a phototransistor. (A phototransistor is much more sensitive than a photodiode.) Solder two wires approximately 15 cm long to the collector and emitter of a phototransistor (the outside leads). Then place a 3 cm piece of heat shrink tubing (approximately 5 mm diameter) over the phototransistor to serve as a light collimator; light should reach the phototransistor only through the tubing as shown in Figure 5-4(a). You may want to *slightly* heat the shrink tube to the transistor.

The phototransistor detector circuit is shown in Figure 5-4(b). Notice that the base lead (center) is left open. The two leads from the phototransistor are brought to the protoboard to complete the circuit. Connect a voltmeter across the collector resistor, R_C.

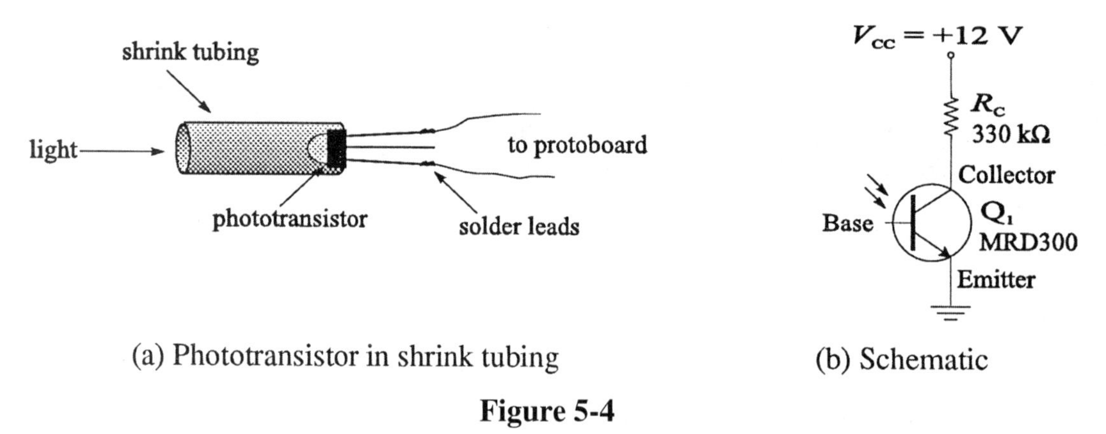

(a) Phototransistor in shrink tubing

(b) Schematic

Figure 5-4

Next, solder two wires approximately 15 cm long to the green LED. Then connect the circuit shown in Figure 5-5. The wires allow the LED to be positioned away from the circuit board. Figure 5-6 is marked for this experiment. Position the lighted LED over the spot marked for it. Tape it down with masking tape. Place the photosensor on the 0° line aimed at the LED. You will need to have room lights dimmed. Record the voltage across the collector resistor. This voltage is proportional to the light from the LED.

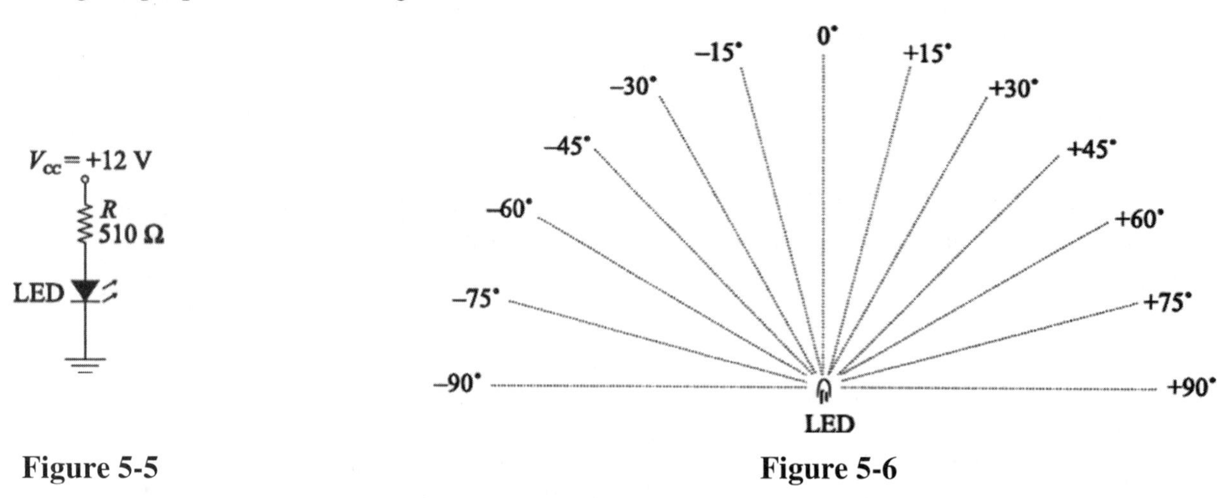

Figure 5-5

Figure 5-6

Keeping the phototransistor at a constant distance from the LED, measure the voltage across the collector resistor, R_C, every 15° around the LED. Set up a data table and record the values. Summarize your findings.

Name ______________
Date ______________
Class ______________

6 Bipolar Junction Transistor Characteristics

Reading:
Floyd and Buchla, *Analog Fundamentals: A Systems Approach*, Section 3-1

Objectives:
After performing this experiment, you will be able to:
1. Measure and graph the collector characteristic curves for a bipolar junction transistor.
2. Use the characteristic curves to determine the β_{DC} of the transistor at a given point.

Summary of Theory:
A bipolar junction transistor (BJT) is a three-terminal device capable of amplifying an ac signal. The three terminals are called the base, emitter, and the collector. BJTs consist of a very thin base material sandwiched in between two of the opposite type materials. They are available in two forms, either *npn* or *pnp*. The middle letter indicates the type of material used for the base, while the outer letters indicate the emitter and collector material. The sandwiched materials produce two *pn* junctions. These two junctions form two diodes – the emitter-base diode and the base-collector diode.

BJTs are current amplifiers. A small base current is amplified to a larger current in the collector-emitter circuit. An important characteristic is the dc current gain, which is the ratio of collector current to base current. This is called the dc beta (β_{DC}) of the transistor. Another useful characteristic is the dc alpha. The dc alpha is the ratio of the collector current to the emitter current and is always less than 1.

For a transistor to amplify, power is required from dc sources. The dc voltages required for proper operation are referred to as bias voltages. The purpose of bias is to establish and maintain the required operating conditions despite variations between transistors or changes in circuit parameters. For normal operation, the base-emitter junction is forward-biased and the base-collector junction is reverse-biased. Since the base-emitter junction is forward-biased, it has characteristics of a forward-biased diode. A silicon bipolar transistor requires approximately 0.7 V of voltage across the base-emitter junction to cause base current.

Materials Needed:
Resistors: One 100 Ω resistor, one 33 kΩ resistor
One 2N3904 *npn* transistor (or equivalent)

For Further Investigation:
Option 1: Transistor curve tracer
Option 2: One rectifier diode (1N4001 or equivalent)
One small power transformer with a 12.6 V ac output

Procedure:

1. Measure and record the resistance of the resistors listed in Table 6-1.

Table 6-1

Resistor	Listed Value	Measured Value
R_1	33 kΩ	
R_2	100 Ω	

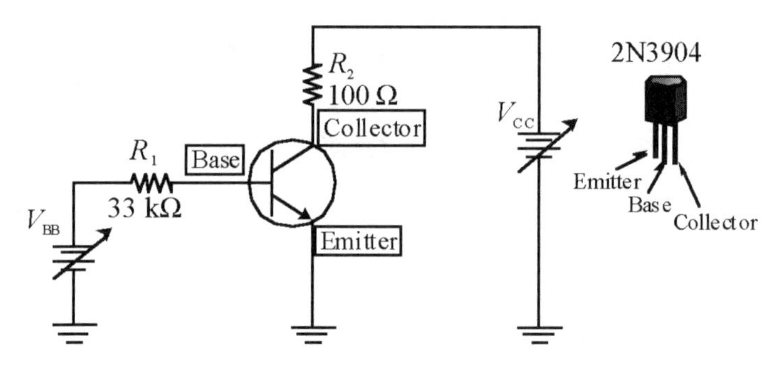

Figure 6-1

2. Connect the common-emitter configuration illustrated in Figure 6-1. Start with both power supplies set to 0 V. The purpose of R_1 is to limit base current to a safe level and to allow indirect determination of the base current. Slowly increase V_{BB} until V_{R1} is 1.65 V. This sets up a base current of 50 μA, which can be shown by applying Ohm's law to R_1.

3. Without disturbing the setting of V_{BB}, slowly increase V_{CC} until +2.0 V is measured between the transistor's collector and emitter. This voltage is V_{CE}. Measure and record V_{R2} for this setting. Record V_{R2} in Table 6-2 in the column labeled Base Current = 50 μA.

Table 6-2

V_{CE}	Base Current = 50 μA		Base Current = 100 μA		Base Current = 150 μA	
(measured)	V_{R2} (measured)	I_C (computed)	V_{R2} (measured)	I_C (computed)	V_{R2} (measured)	I_C (computed)
2.0 V						
4.0 V						
6.0 V						
8.0 V						

4. Compute the collector current, I_C, by applying Ohm's law to R_2. Use the measured voltage, V_{R2}, and the measured resistance, R_2, to determine the current. Note that the current in R_2 is the same as I_C for the transistor. Enter the computed collector current in Table 6-2 in the column labeled Base Current = 50 μA.

5. Without disturbing the setting of V_{BB}, increase V_{CC} until 4.0 V is measured across the transistor's collector to emitter. Measure and record V_{R2} for this setting. Compute the collector current by applying Ohm's law as in step 4. Continue in this manner for each of the values of V_{CE} listed in Table 6-2.

6. Reset V_{CC} for 0 V and adjust V_{BB} until V_{R1} is 3.3 V. The base current is now 100 μA.

7. Without disturbing the setting of V_{BB}, slowly increase V_{CC} until V_{CE} is 2.0 V. Measure and record V_{R2} for this setting in Table 6-2 in the column labeled Base Current = 100 μA. Compute I_C for this setting by applying Ohm's law to R_2. Enter the computed collector current in Table 6-2.

8. Increase V_{CC} until V_{CE} is equal to 4.0 V. Measure and record V_{R2} for this setting. Compute I_C as before. Continue in this manner for each value of V_{CC} listed in Table 6-2.

9. Reset V_{CC} for 0 V and adjust V_{BB} until V_{R1} is 4.95 V. The base current is now 150 μA.

10. Complete Table 6-2 by repeating steps 7 and 8 for 150 μA of base current.

11. Plot three collector characteristic curves using the data tabulated in Table 6-2. The collector characteristic curve is a graph of V_{CE} versus I_C for a constant base current. Choose a scale for I_C that allows the largest current observed to fit on the graph. Label each curve with the base current it represents. Graph the data on Plot 6-1 below.

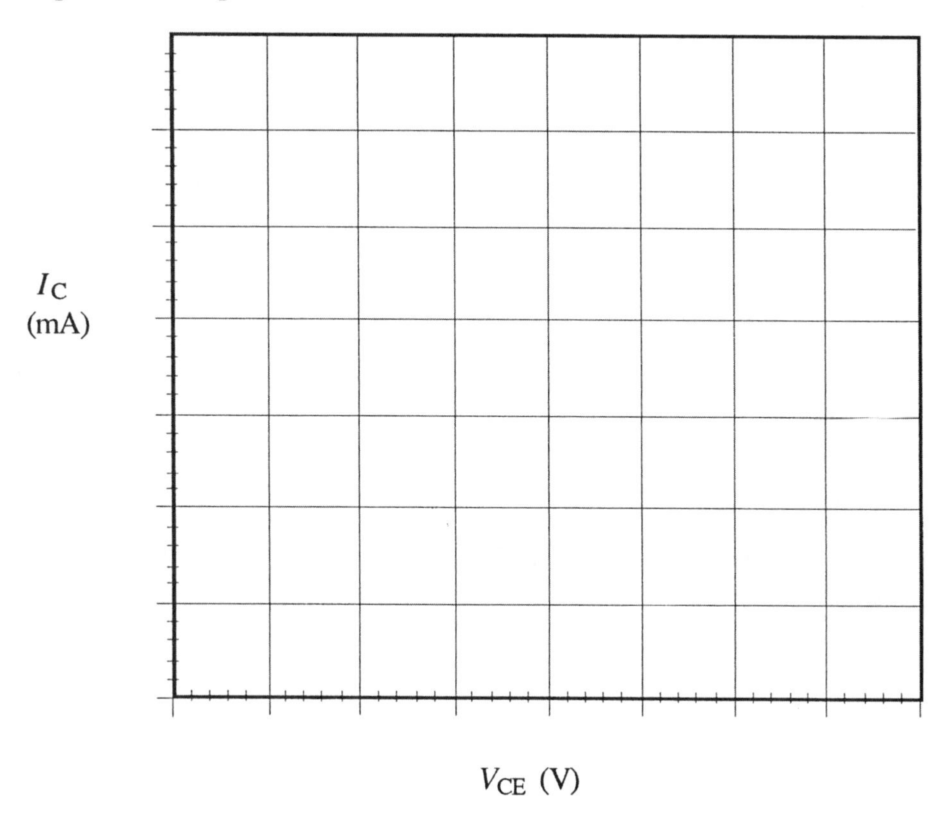

Plot 6-1

12. Use the characteristic curve you plotted to determine the β_{DC} for the transistor at a V_{CE} of 3.0 V and a base current of 50 μA, 100 μA, and 150 μA. Then repeat the procedure for a β_{DC} at a V_{CE} of 5.0 V. Tabulate your results in Table 6-3.

Table 6-3

	Current Gain, β_{DC}		
V_{CE}	I_B = 50 μA	I_B = 100 μA	I_B = 150 μA
3.0 V			
5.0 V			

Conclusion:

Evaluation and Review Questions:

1. Does the experimental data indicate that β_{DC} is a constant at all points? Does this have any effect on the linearity of the transistor?

2. What effect would a higher β_{DC} have on the characteristic curves you measured?

3. What is the maximum power dissipated in the transistor for the data taken in the experiment?

4. (a) The dc alpha of a bipolar transistor is the collector current, I_C, divided by the emitter current, I_E. Using this definition and $I_E = I_C + I_B$, show that dc alpha can be written as:

$$\alpha_{DC} = \frac{\beta_{DC}}{\beta_{DC} + 1}$$

   (b) Compute dc alpha for your transistor at $V_{CE} = 4.0$ V and $I_B = 100$ μA.

5. What value of V_{CE} would you expect if the base terminal of a transistor were open? Explain your answer.

For Further Investigation:

Option 1:

If you have a transistor curve tracer available, you can use it to check the data taken in this experiment. A transistor curve tracer has a step generator that generates a staircase set of current or voltage steps. Set the step generator to 50 μA per step. Select positive steps to apply to the base with the emitter grounded. Select a positive sweep voltage of approximately +20 V with a series limiting resistance of several hundred ohms. Select a horizontal display of 1 V/div and a vertical display of about 10 mA/div. (If your transistor has a very high or low $ß_{DC}$, you may need to change these settings.) The curve tracer will show the collector characteristic curves. Test the effect of heating or cooling the transistor on the $ß_{DC}$.

Option 2:

If you do not have a transistor curve tracer available, you can still observe the collector curves, one at a time, on an oscilloscope. The circuit is a modification of the one used in the experiment and is shown in Figure 6-2. The collector supply is replaced with a low voltage transformer and diode. Start with V_{BB} set to 1.65 V as before (I_B = 50 μA).

Put the oscilloscope in X-Y mode and put both channels to the GND (ground) position. Keep the intensity low and position the dot in the lower left corner of the screen. Adjust the oscilloscope Y channel to 0.1 V/div (equivalent to 1.0 mA/div) and the X channel to 1 V/div and invert the Y channel.[1] Couple the signal to the scope and raise the intensity; you should see the first collector curve that you measured in the experiment. You can adjust V_{BB} to observe the other curves.

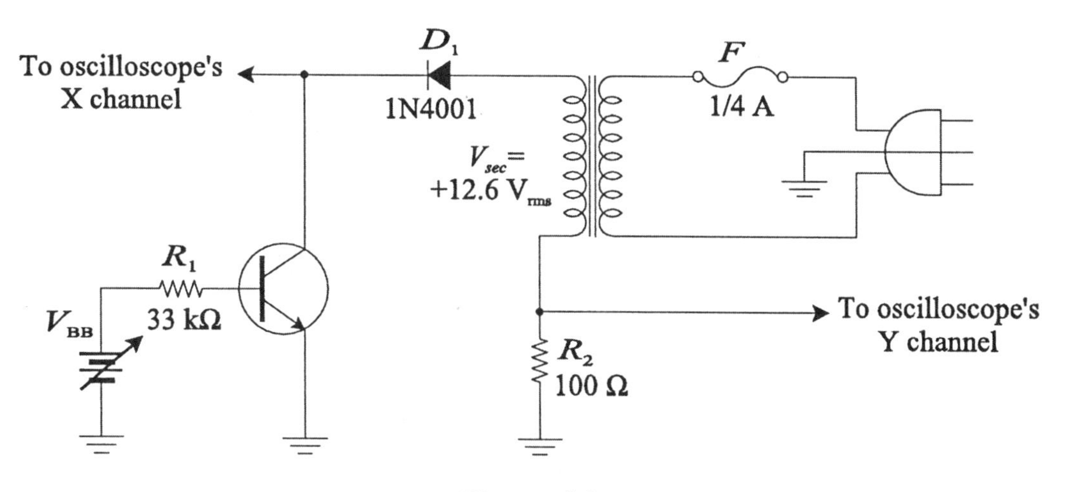

Figure 6-2

[1] If you cannot invert the Y channel, position the trace at the top of the screen. Increasing current will be toward the bottom of the screen.

Name ______________
Date ______________
Class ______________

7 Bipolar Transistor Biasing

Reading:
Floyd and Buchla, *Analog Fundamentals: A Systems Approach*, Section 3-2

Objectives:
After performing this experiment, you will be able to:
1. Construct and analyze three types of transistor bias circuits: base bias, voltage-divider bias, and collector-feedback bias. Compare the stability of the bias with different transistors.
2. Select appropriate bias resistors for each type of bias.
3. *For Further Investigation:* Construct and analyze an emitter-bias circuit.

Summary of Theory:
For a transistor to amplify signals, it is necessary to forward-bias the base-emitter junction and to reverse-bias the base-collector junction. The purpose of bias is to provide dc voltages to set up the proper quiescent (no signal) conditions for circuit operation.

There are four common bias circuits for bipolar transistors. You should be familiar with the advantages and disadvantages of each. The four circuits are: (1) base bias, (2) emitter bias, (3) voltage-divider bias, and (4) collector-feedback bias. These basic bias circuits are illustrated in Figure 7-1 for *npn* transistors. These bias methods apply to *pnp* transistors by reversing voltage polarities. The key for either type of transistor is that the base-emitter junction is forward-biased and the base-collector junction is reverse-biased.

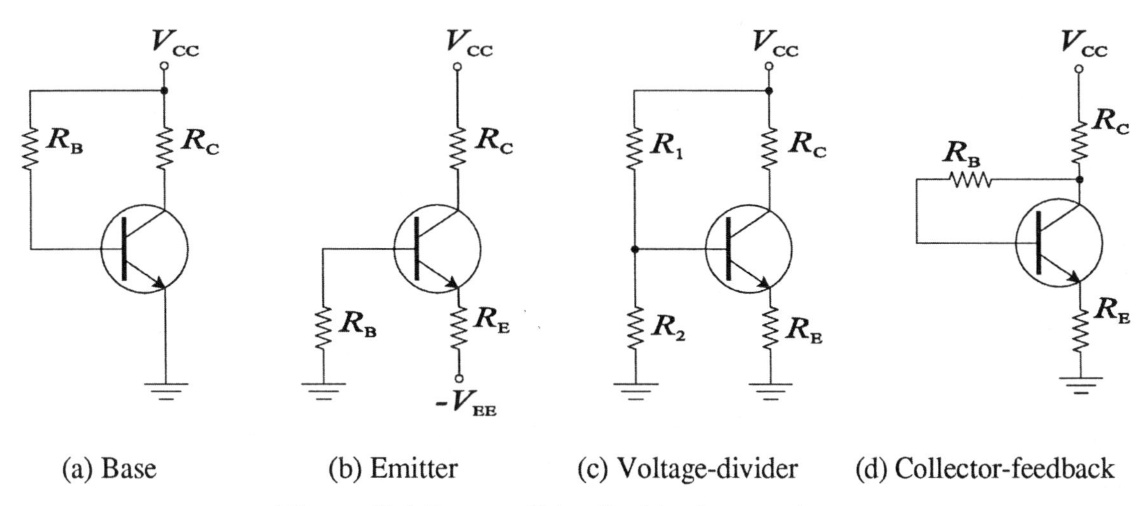

Figure 7-1 Types of bias for bipolar transistors.

Base bias is the simplest form because it uses a single power supply and resistor. It is satisfactory for switching applications but is generally unsatisfactory for linear circuits due to ß dependency. Normal variations in transistors greatly affect the operating point of base-biased circuits.

Emitter bias overcomes the difficulty of ß dependency, but requires a positive and a negative power supply. The dc conditions can be found by writing Kirchhoff's Voltage Law (KVL) around the base-emitter circuit and solving for the emitter current. The emitter current is approximately equal to the collector current, and voltages can be found by applying Ohm's law. Emitter bias is discussed in Section 3-2 of the text.

Voltage-divider bias is widely used because it is stable yet requires only one power supply. When the divider current is much larger than the base current, the small base current can be ignored, simplifying the analysis. This is called "stiff" bias. The steps to solve for the dc parameters for the CE amplifier with stiff voltage-divider bias are given as follows:

1. Mentally remove the capacitors from the circuit since they appear open to dc. For this circuit, this causes the load resistor, R_L, to be removed (see Figure 7-2(a)).
2. Solve for the base voltage, V_B, by applying the voltage-divider rule to R_1 and R_2 as illustrated in Figure 7-2(b).
3. Subtract the 0.7 V forward-bias drop across the base-emitter diode from V_B to obtain the emitter voltage, V_E, as illustrated in Figure 7-2(c).
4. The dc current in the emitter circuit is found by applying Ohm's law to R_E. The emitter current, I_E, is approximately equal to the collector current, I_C. The transistor appears to be a current source of approximately I_E into the collector circuit as shown in Figure 7-2(d).

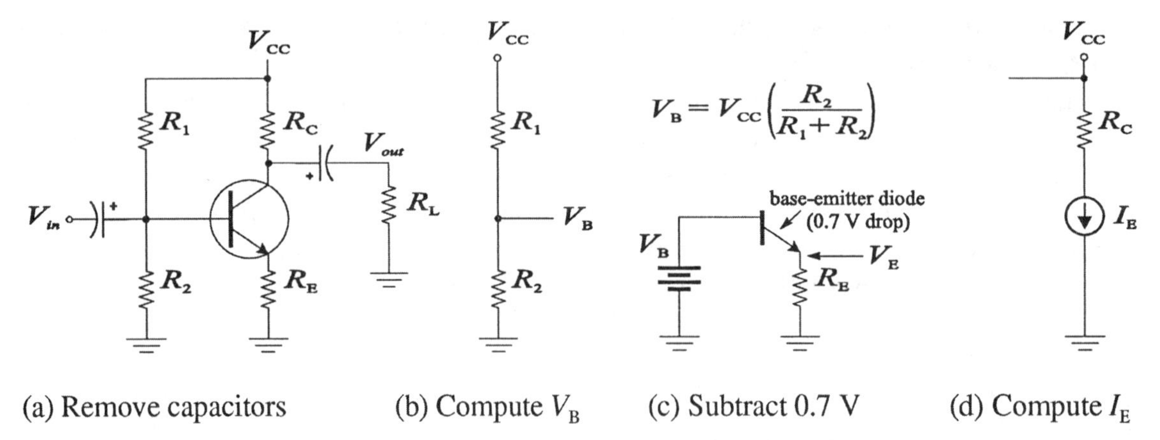

(a) Remove capacitors (b) Compute V_B (c) Subtract 0.7 V (d) Compute I_E

Figure 7-2 Steps in solving a CE amplifier with stiff voltage-divider bias.

Collector-feedback bias uses a form of negative feedback to stabilize the Q-point. The analysis of collector-feedback bias can be done by writing KVL through the base circuit and finding the emitter (or collector) current. Note that for this form of biasing, the *collector* resistor actually has the *emitter* current! The derivation is given in step 8 of the Procedure.

Materials Needed:

Resistors (one of each): 470 Ω, 2.0 kΩ, 6.8 kΩ, 33 kΩ, 360 kΩ, 1.0 MΩ
Three small signal *npn* transistors, (2N3904 or equivalent)

For Further Investigation:

Resistors: one 3.6 kΩ, one 100 kΩ

Procedure:

1. Measure and record the values of the resistors listed in Table 7-1.

Table 7-1

Resistor	Listed Value	Measured Value
R_B	1.0 MΩ	
R_C	2.0 kΩ	

2. You will test each of the three transistors, one at a time, in a base bias circuit. The manufacturer's specification sheet for a 2N3904 transistor indicates that β_{DC} can range from 100 to 400. Assuming the β_{DC} is 200, compute the parameters listed in Table 7-2 for the base bias circuit shown in Figure 7-3. Start by computing the voltage across the base resistor, V_{RB}, and the current in this resistor, I_B. Using β_{DC} find the collector current, I_C, the voltage across the collector resistor, V_{RC}, and the voltage from collector to ground, V_C.

Table 7-2

DC Parameter	Computed Value	Measured Value Q_1	Q_2	Q_3
V_{RB}				
I_B				
I_C				
V_{RC}				
V_C				

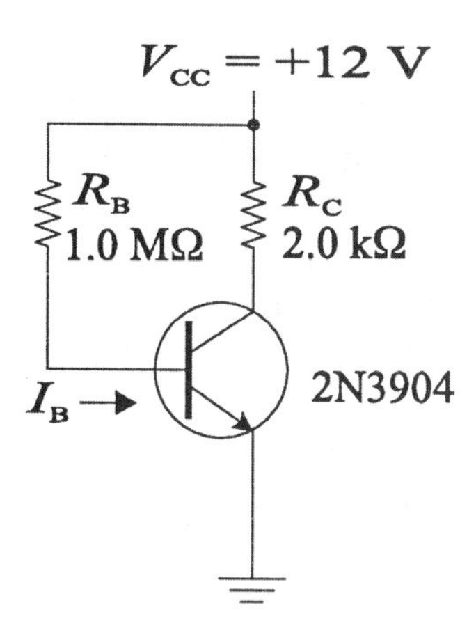

Figure 7-3

3. Label each of three *npn* transistors as Q_1, Q_2, and Q_3. Construct the circuit shown in Figure 7-3 using Q_1. Measure the voltages listed in Table 7-2 for Q_1. Then remove Q_1 from the circuit and test the other two transistors in the same circuit. Record all measurements in Table 7-2.

4. Test voltage-divider bias for the same three transistors. Start by measuring and recording the values of the resistors listed in Table 7-3.

Table 7-3

Resistor	Listed Value	Measured Value
R_1	33 kΩ	
R_2	6.8 kΩ	
R_E	470 Ω	
R_C	2.0 kΩ	

5. Compute the parameters listed in Table 7-4 for the circuit shown in Figure 7-4. The method is outlined in the box shown in the Summary of Theory. Note that the bias is relatively "stiff" so the approximations given in the Summary of Theory are reasonable.

Table 7-4

DC Parameter	Computed Value	Measured Value Q_1	Q_2	Q_3
V_B				
V_E				
$I_E \approx I_C$				
V_{RC}				
V_C				

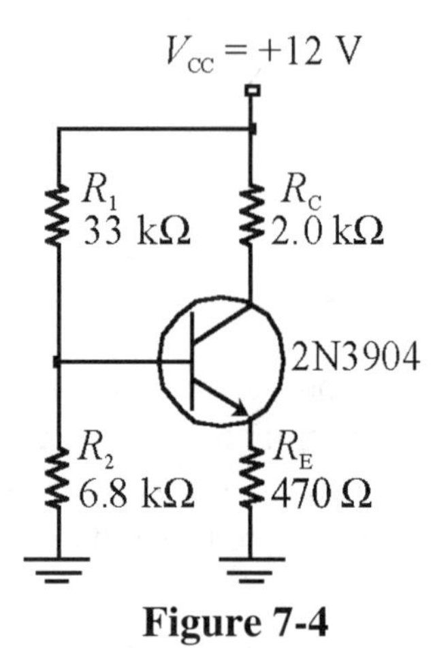

Figure 7-4

6. Construct the circuit shown in Figure 7-4 using transistor Q_1. Measure the voltages listed in Table 7-4 for Q_1. Then remove Q_1 from the circuit and test the other two transistors in the same circuit. Record all measurements in Table 7-4.

7. In this step, you will compare the same three transistors in a collector-feedback circuit. Measure and record the values of the resistors listed in Table 7-5.

Table 7-5

Resistor	Listed Value	Measured Value
R_B	360 kΩ	
R_C	2.0 kΩ	

8. Compute the parameters listed in Table 7-6 for the circuit shown in Figure 7-5. To find the approximate collector and emitter currents, you can write Kirchhoff's voltage law around the base path as follows:

$$-V_{CC} + I_E R_C + I_B R_B + V_{BE} = 0$$

Substituting for I_B, $I_B \cong \dfrac{I_E}{\beta_{DC}}$

And solving for I_E, we obtain: $I_E \cong I_C = \dfrac{V_{CC} - V_{BE}}{R_C + \dfrac{R_B}{\beta_{DC}}}$

Assume the β_{DC} is 200 for the calculation. Then find the voltage across the collector resistor, V_{RC}, and the collector voltage, V_C.

9. Construct the circuit shown in Figure 7-5 using transistor Q_1. Measure the voltages listed in Table 7-6 for Q_1. Then remove Q_1 from the circuit and test the other two transistors in the same circuit. Record all measurements in Table 7-6.

Table 7-6

DC Parameter	Computed Value	Measured Value Q_1	Q_2	Q_3
I_C				
V_{RC}				
V_C				

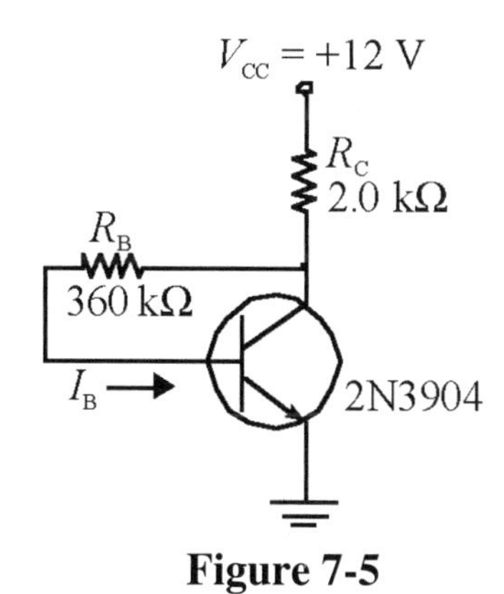

Figure 7-5

Conclusion:

Evaluation and Review Questions:

1. Compare your observations of the three bias methods tested in the experiment. Which showed the *least* variation between the transistors?

2. Draw each of the three bias circuits tested in the experiment for the case of a *pnp* transistor. Assuming the same resistors as used in the experiment, compute the base, emitter and collector voltages. Show these voltages on your drawing.

   (a) base bias (b) voltage-divider bias (c) collector-feedback bias

3. Assume you need to bias the amplifier shown in Figure 7-6. It is desired to have the Q-point set to approximately 20 mA of collector current. Choose bias resistors for voltage-divider bias that will provide reasonably stiff bias. Show your work.

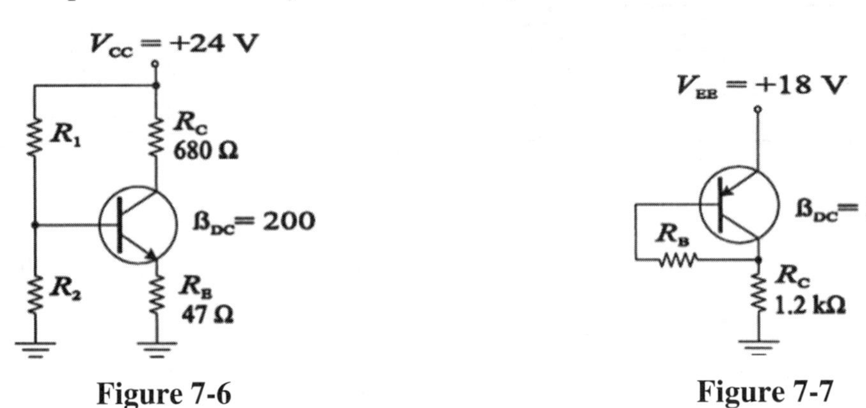

Figure 7-6 **Figure 7-7**

4. Assume you need to bias the amplifier shown in Figure 7-7. It is desired to have the Q-point set to approximately 5.0 mA of collector current. Choose a base resistor for collector-feedback bias that will provide reasonably stiff bias. Show your work.

5. For the circuit in Figure 7-6, predict the effect of each of the following problems on the collector voltage:

(a) R_1 opens

(b) Base is shorted to ground through a solder bridge

(c) R_E is 470 Ω instead of 47 Ω

(d) V_{CC} drops to +15 V

For Further Investigation:

As discussed in the Summary of Theory, emitter bias is an excellent way of obtaining stable bias, however, it requires both a positive and negative power supply. Figure 7-8 shows a transistor with emitter bias. Compute the dc parameters for the circuit, then build and test it. Compare the measured dc parameters for each of the three transistors you used in this experiment. Summarize your findings in a short report.

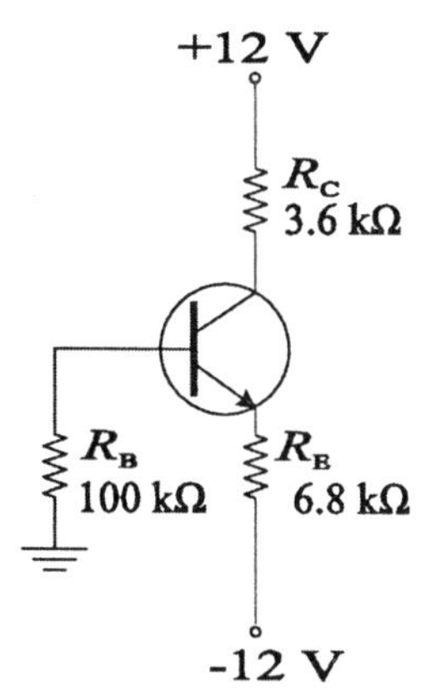

Figure 7-8

8 The Common-Emitter Amplifier	Name ______________ Date ______________ Class ______________

Reading:
Floyd and Buchla, *Analog Fundamentals: A Systems Approach*, Section 3-4

Objectives:
After performing this experiment, you will be able to:

1. Compute the dc and ac parameters for a common-emitter (CE) amplifier.
2. Build a CE amplifier circuit and measure the dc parameters, the ac input resistance, and the voltage gain. Observe the phase relationship between the input and output signals.
3. Predict and test the effects of certain faults in a CE amplifier.

Summary of Theory:
In a common-emitter (CE) amplifier, the input signal is applied between the base and emitter and output signal is developed between the collector and emitter. The transistor's *emitter* is common to both the input and output circuits, hence, the term *common emitter*. A CE amplifier is illustrated in Figure 8-1(a). This is the basic circuit that will be tested in this experiment. It will be used again in the Further Investigation of Experiment 9 to drive a common-collector (CC) amplifier.

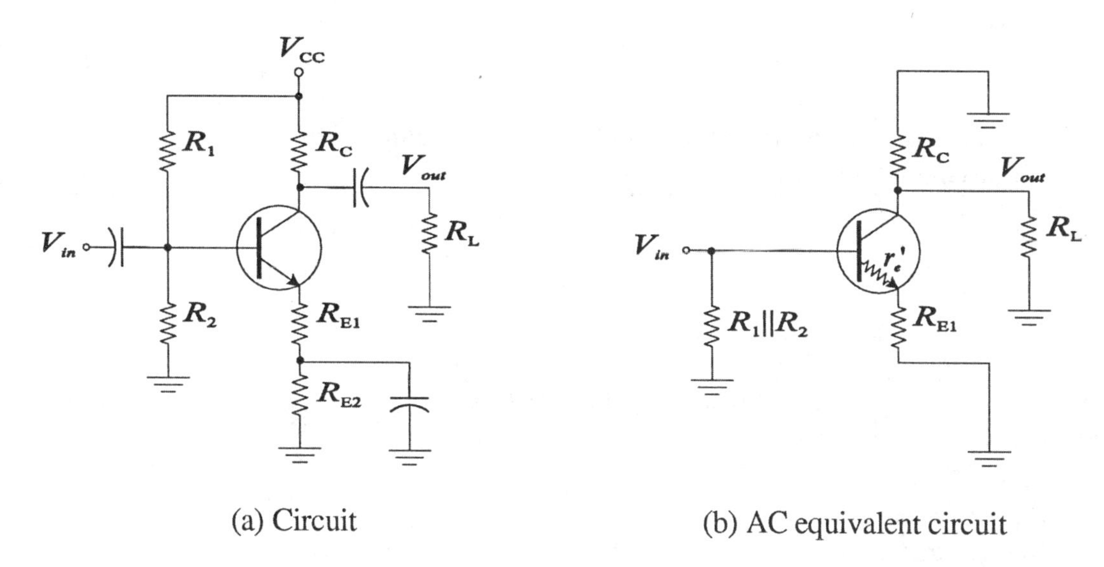

(a) Circuit

(b) AC equivalent circuit

Figure 8-1

To amplify ac signals, the base-emitter junction must be forward-biased and the base-collector junction must be reverse-biased. The bias establishes and maintains the proper dc operating conditions for the transistor. The dc parameters are analyzed first as described in Experiment 7 and the text.

After analyzing the dc conditions, the ac parameters for the amplifier can be evaluated. The equivalent ac circuit is drawn in Figure 8-1(b). The capacitors appear to be an ac short. Thus, the ac equivalent circuit does not contain R_{E2} in this example. Using the superposition theorem, V_{CC} is replaced with a short, placing it at ac ground. The analysis steps are:

1. Replace all capacitors with a short and place V_{CC} at ac ground. Compute the ac resistance of the emitter, r_e', from the equation:

$$r_e' = \frac{25\text{ mV}}{I_E}$$

2. Compute the amplifier's voltage gain. Voltage gain is the ratio of the output voltage divided by the input voltage. The input voltage is across the ac emitter resistance to ground which, in this example, is $r_e' + R_{E1}$. The output voltage is taken across the ac resistance from collector to ground. Looking from the transistor's collector, R_L appears to be in parallel with R_C. Also, I_c is approximately equal to I_e. For the circuit in Figure 8-1(b), the output voltage divided by the input voltage can be written:

$$A_v = \frac{V_{out}}{V_{in}} = \frac{I_c\left(R_C \parallel R_L\right)}{I_e\left(r_e' + R_{E1}\right)} \cong \frac{\left(R_C \parallel R_L\right)}{\left(r_e' + R_{E1}\right)}$$

3. Compute the total input resistance seen by the ac signal:

$$R_{in\,(tot)} = R_1 \parallel R_2 \parallel \beta_{ac}\left(r_e' + R_{E1}\right)$$

Notice that the ac resistance of the emitter circuit is multiplied by β_{ac} when it is brought into the base circuit.

Materials Needed:

Resistors:
one 100 Ω, one 330 Ω, two 1.0 kΩ, one 4.7 kΩ, two 10 kΩ
Capacitors:
two 1.0 μF, one 47 μF
One 10 kΩ potentiometer
One 2N3904 *npn* transistor (or equivalent)

Procedure:

1. Measure and record the resistance of the resistors listed in Table 8-1.

Table 8-1

Resistor	Listed Value	Measured Value
R_1	10 kΩ	
R_2	4.7 kΩ	
R_{E1}	100 Ω	
R_{E2}	330 Ω	
R_C	1.0 kΩ	
R_L	10 kΩ	

Table 8-2

DC Parameter	Computed Value	Measured Value
V_B		
V_E		
I_E		
V_C		
V_{CE}		

2. Compute the dc parameters listed in Table 8-2 for the CE amplifier shown in Figure 8-2. (Review Experiment 7 for the method.) Note that V_B, V_E, and V_C are with respect to circuit ground. Use the sum of R_{E1} and R_{E2} times I_E to compute the dc emitter voltage, V_E. Compute V_C by subtracting V_{RC} from V_{CC}. Enter your computed values in Table 8-2.

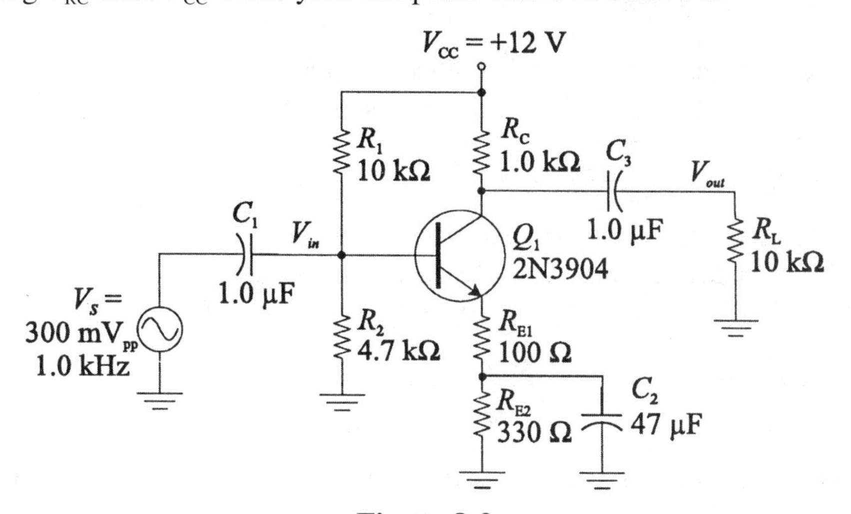

Figure 8-2

3. Construct the amplifier shown in Figure 8-2. The signal generator should be turned off. Measure and record the dc voltages listed in Table 8-2.

4. Compute the ac parameters listed in Table 8-3. The input signal, V_{in}, is set for 300 mV_{PP}. This is both V_{in} and the ac base voltage, V_b. Multiply V_{in} by the computed voltage gain to calculate the ac voltage at the collector; this is both V_c and V_{out}. If you do not know the $ß_{ac}$ for the input resistance calculation, assume a value of 100.

5. Turn on the signal generator and set V_{in} for 300 mV_{PP} at 1.0 kHz with the generator connected to the circuit. Use the oscilloscope to set the proper voltage and check the frequency. Measure the ac signal voltage at the transistor's emitter and at the collector. Note that the signal at the emitter is less than the base (why?). Use V_{in} and the ac collector voltage (V_{out}) to determine the measured voltage gain, A_v. The measurement of $R_{in(tot)}$ is explained in step 6. Record the ac measurements in Table 8-3.

Table 8-3

AC Parameter	Computed Value	Measured Value
$V_{in} = V_b$	300 mV_{pp}	
V_e		
r_e'		
A_v		
$V_{out} = V_c$		
$R_{in(tot)}$		

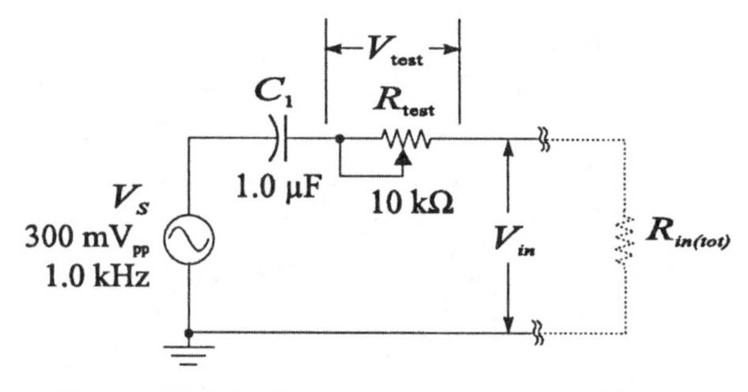

Figure 8-3 Indirect measurement of $R_{in(tot)}$.

6. The measurement of $R_{in(tot)}$ is done indirectly since it is an ac resistance that cannot be measured with an ohmmeter. The output signal (V_{out}) is measured with an oscilloscope and recorded with the amplifier operating normally (no clipping or distortion). A rheostat (R_{test}) is then inserted in series with the source as shown in Figure 8-3. The rheostat is varied until V_{out} drops to one-half the value prior to inserting R_{test}. With this condition, $V_{in} = V_{test}$ and $R_{in(tot)}$ must be equal to R_{test}. R_{test} can then be removed and measured with an ohmmeter. Using this method, measure $R_{in(tot)}$ and record the result in Table 8-3.

7. Restore the circuit to that of Figure 8-2. With a two-channel oscilloscope, compare the input and output waveforms. What is the phase relationship between V_{in} and V_{out}?

____

8. Remove the bypass capacitor, C_2, from the circuit. Measure the ac signal voltage at the transistor's base, emitter, and collector. Measure the voltage gain of the amplifier. What conclusion can you make about the amplifier's performance with C_2 open?

____

9. Replace C_2 and reduce R_L to 1.0 kΩ. Observe the ac signal voltage at the transistor's base, emitter, and collector and measure the voltage gain of the amplifier. What conclusion can you make about the amplifier's performance with R_L reduced to 1.0 kΩ?

____

10. Replace R_L with the original 10 kΩ resistor and open R_{E1}. Measure the dc voltages at the base, emitter, and collector. Is the transistor in cutoff or in saturation? Explain.

____

11. Replace R_{E1} and open R_2. Measure the dc voltages at the base, emitter and collector. Is the transistor in cutoff or saturation? Explain.

____

Note: The amplifier from this experiment is used again in the For Further Investigation section of Experiment 9. You may want to save the circuit from this experiment.

Conclusion:

Evaluation and Review Questions:

1. When the bypass capacitor, C_2, is open, you found that the gain is affected. Explain.

2. In step 6, you were instructed to measure the input resistance while monitoring the output voltage. Why is this procedure better than monitoring the base voltage?

3. Assume the amplifier shown in Figure 8-2 has +1.8 V dc measured on the base, +1.1 V dc measured on the emitter, and +1.1 V dc measured on the collector.

   (a) Is this normal?

   (b) If not, what is the most likely cause of the problem?

4. If C_2 were shorted,

   (a) what dc base voltage would you expect? ________________

   (b) what dc collector voltage would you expect? ________________

5. Explain a simple test to determine if a transistor is in saturation or in cutoff.

For Further Investigation:

The low frequency response of the CE amplifier in this experiment is determined by the coupling and bypass capacitors, C_3 and C_2, respectively. The upper frequency response is determined by the unseen interelectrode and stray circuit capacitances. Using the oscilloscope to view the output waveform, set the generator for a midband frequency of 1.0 kHz. Use a sine wave with a convenient level (not clipped) across the load resistor. Raise the generator frequency until the output voltage falls to 70.7% of the midband level. This is the upper cutoff frequency. Then lower the generator frequency until the output voltage falls to 70.7% of the midband level. This is the lower cutoff frequency. Try switching C_1 with C_2 in the circuit of Figure 8-2. What effect does this have on the lower cutoff frequency? Does it have an effect on the upper cutoff frequency? Summarize your investigation in a short report.

Name ____________
Date ____________
Class ____________

9 The Common-Collector Amplifier

Reading:
Floyd and Buchla, *Analog Fundamentals: A Systems Approach*, Section 3-5

Objectives:
After performing this experiment, you will be able to:
1. Compute the dc and ac parameters for a common-collector amplifier.
2. Build the amplifier from objective 1 and measure the dc and ac parameters including input resistance and power gain.
3. Test the effect of different load resistors on the ac parameters.
4. Predict the effect of faults in a common-collector amplifier.

Summary of Theory:
The common-collector (CC) amplifier (also called the *emitter-follower*) has the input signal applied to the base and the output signal is taken from the emitter. Figure 9-1(a) illustrates a CC amplifier using a *pnp* transistor with voltage-divider bias. The ac output voltage almost perfectly duplicates the input voltage waveform. While this implies that the voltage gain is approximately 1, the current gain is not; hence, it can deliver increased signal power to a load. The CC amplifier is characterized by a high input resistance and a low output resistance.

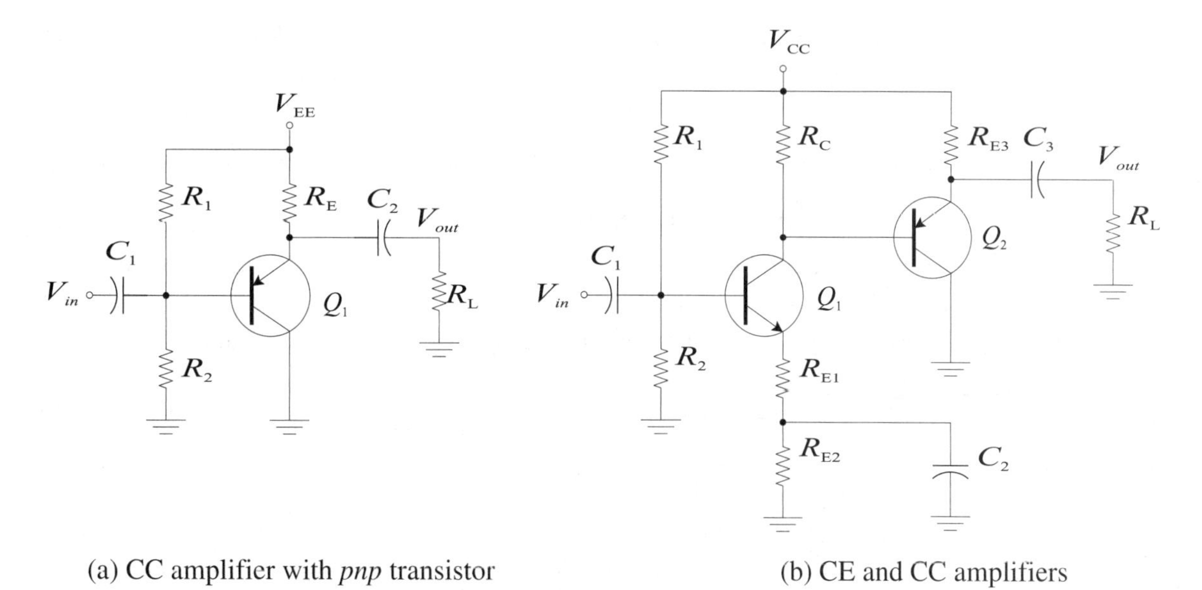

(a) CC amplifier with *pnp* transistor

(b) CE and CC amplifiers

Figure 9-1

Frequently, a CC amplifier follows a voltage amplifier. Instead of having separate bias resistors, bias may be obtained through a dc path connected from the previous stage as illustrated in Figure 9-1(b). This technique is common in power amplifiers. It will be investigated in the For Further Investigation for this experiment.

Analysis of the amplifier begins with the **dc** parameters. These procedures are described in Experiment 7 and summarized here for the *pnp* transistor shown in Figure 9-1(a).

1. Mentally remove the capacitors from the circuit since they appear open to dc. This causes the load resistor, R_L, to be removed.
2. Solve for the base voltage, V_B, by applying the voltage-divider rule to R_1 and R_2.
3. *Add* the 0.7 V forward-bias drop across the base-emitter diode from V_B to obtain the emitter voltage, V_E. (Note that the emitter is at a higher voltage in the *pnp* case.)
4. The dc current in the emitter circuit is found by applying Ohm's law to R_E. The voltage across the emitter resistor is the difference between the supply voltage (V_{EE}) and V_E. The collector current is nearly equal to the emitter current, and the collector voltage is zero.

The **ac** parameters for the amplifier can now be analyzed. The equivalent ac circuit is illustrated in Figure 9-2. The analysis steps are:

1. Replace all capacitors with a short. Compute the ac resistance of the emitter, r_e', from the equation:

$$r_e' = \frac{25\text{ mV}}{I_E}$$

2. Compute the amplifier's voltage gain. Voltage gain is the ratio of the output voltage divided by the input voltage. The input voltage is applied across r_e' and the ac emitter resistance, whereas the output voltage is taken only across the ac emitter resistance. Thus, the voltage gain is based on the voltage divider equation:

$$A_v = \frac{V_{out}}{V_{in}} = \frac{I_c(R_E \parallel R_L)}{I_e(r_e' + R_E \parallel R_L)} = \frac{R_E \parallel R_L}{r_e' + R_E \parallel R_L}$$

3. Compute the total input resistance seen by the ac signal:

$$R_{in\ (tot)} = R_1 \parallel R_2 \parallel \{\beta_{ac}(r_e' + R_E \parallel R_L)\}$$

4. Compute the amplifier's power gain. In this case, we are only interested in the power delivered to the load resistor. The output power is V_{out}^2 / R_L. The input power is $V_{in}^2 / R_{in(tot)}$. Since the voltage gain is approximately 1, the power gain can be expressed as a ratio of $R_{in(tot)}$ to R_L:

$$A_p = \frac{\dfrac{V_{out}^2}{R_L}}{\dfrac{V_{in}^2}{R_{in\ (tot)}}} = A_v^2 \frac{R_{in\ (tot)}}{R_L} = \frac{R_{in\ (tot)}}{R_L}$$

These formulas were derived for the particular CC amplifier given in the example. You should not assume that these equations are valid for other configurations.

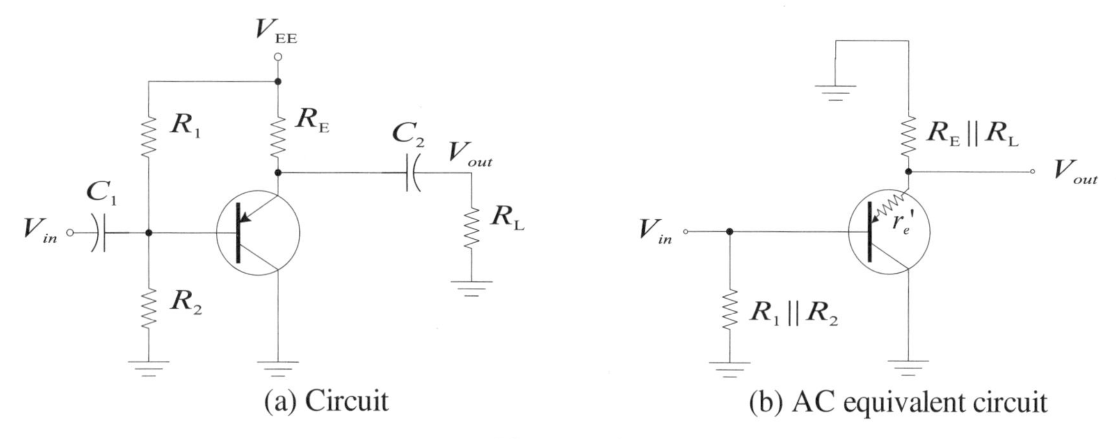

Figure 9-2

Materials Needed:

Resistors:
  two 1.0 kΩ, one 10 kΩ, one 33 kΩ

Capacitors:
  one 1.0 μF, one 10 μF

One 10 kΩ potentiometer

One 2N3906 *pnp* transistor (or equivalent)

For Further Investigation:
  One 2N3906 *pnp* transistor (or equivalent)
  Three 330 Ω resistors

Procedure:

1. Measure and record the resistance of the resistors listed in Table 9-1.

Table 9-1

Resistor	Listed Value	Measured Value
R_1	33 kΩ	
R_2	10 kΩ	
R_E	1.0 kΩ	
R_L	1.0 kΩ	

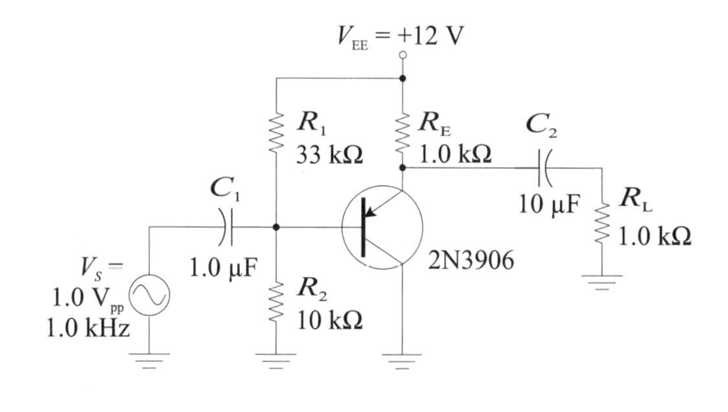

Figure 9-3

2. Compute the dc parameters listed in Table 9-2 for the CC amplifier shown in Figure 9-3. (See Summary of Theory for method.) Enter your computed values in Table 9-2.

Table 9-2

DC Parameter	Computed Value	Measured Value
V_B		
V_E		
I_E		
V_{CE}		

Table 9-3

AC Parameter	Computed Value	Measured Value
V_b	1.0 V_{pp}	
V_e		
r_e'		
A_v		
$R_{in(tot)}$		
A_p		

3. Construct the amplifier shown in Figure 9-3. The signal generator should be turned off. With the power supply on, measure and record the dc voltages listed in Table 9-2. Your measured and computed values should agree within 10%.

4. Compute the ac parameters listed in Table 9-3. Assume V_b is the same as the source voltage, V_s. If you do not know the ß for your transistor, assume it is 100. Use the procedure outlined in the Summary of Theory to compute the parameters.

5. Turn on the signal generator and set V_s for 1.0 V_{pp} at 1.0 kHz. Use the oscilloscope to set the proper voltage and check the frequency. Measure the ac signal voltage at the transistor's emitter, V_{out}, and determine the voltage gain, A_v. Measure $R_{in(tot)}$ using the method employed for the CE amplifier in Experiment 8. Use the measured $R_{in(tot)}$ and R_L to determine the measured power gain.

6. With a two-channel oscilloscope, compare the input and output waveforms. What is the phase relationship between V_{in} and V_{out}?

   ____

7. Table 9-4 lists some possible troubles with the CC amplifier. For each trouble listed, predict the effect on the dc voltages. Then insert the trouble into the circuit and test your prediction. Insert the open collector and open emitter troubles by removing the transistor lead and measuring the voltages at the circuit. For each fault, describe the effect on the ac output waveform (clipped, no output, etc.).

Table 9-4

Trouble	DC Predictions			DC Measurements			Effect of Trouble on V_{out}
	V_B	V_E	V_{CE}	V_B	V_E	V_{CE}	
R_1 open							
R_2 open							
R_1 shorted							
R_E open							
open collector							
open emitter							

8. Replace R_L with a 10 kΩ variable resistor set to 1.0 kΩ. Connect an oscilloscope probe to the emitter. Raise the signal amplitude until you just begin to observe clipping. If the positive peaks are clipped, you are observing *cutoff* clipping because the transistor is turned off. If the negative peaks are clipped, this is called *saturation* clipping because the transistor is fully conducting. What type of clipping is first observed?

   ____

9. Vary R_L while observing the output waveform. Describe your observations.

   ____

   ____

Conclusion:

Evaluation and Review Questions:

1. In step 6, you observed the phase relationship between the input and output waveforms. Is the phase relationship you observed the same for an *npn* circuit? Explain.

2. In step 8, you observed the effect of clipping due to saturation or cutoff of the transistor. The statement was made that if the positive peaks are clipped, you are observing *cutoff* clipping because the transistor is turned off. Is this statement true if the CC circuit had been constructed with an *npn* transistor? Why or why not?

3. The circuit used in this experiment used voltage-divider bias.
   (a) Compared to base bias, what is the advantage?

   (b) What disadvantage does it have?

4. Common-collector amplifiers do not have voltage gain but still provide power gain. Explain.

5. Figure 9-4 shows a CC amplifier with voltage-divider bias. Assume $\beta_{ac} = \beta_{dc} = 100$. Compute the dc and ac parameters listed below for the circuit.

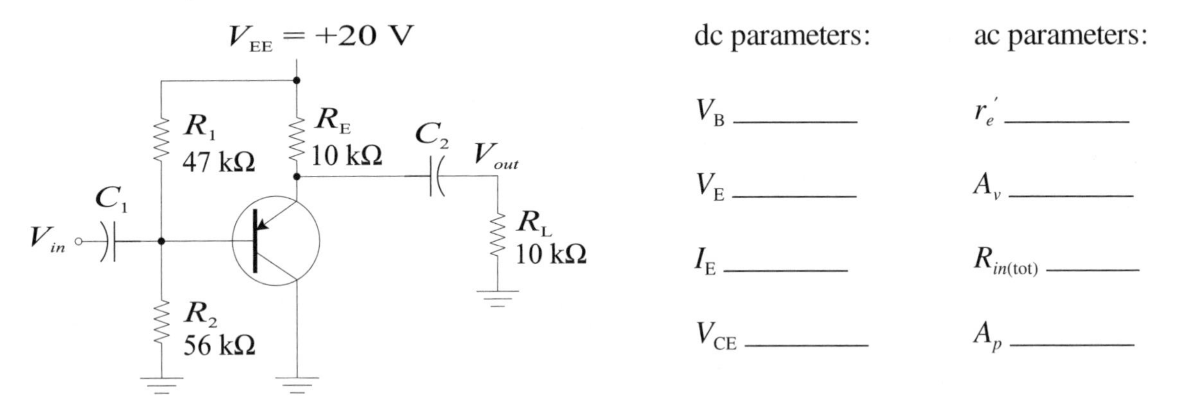

Figure 9-4

For Further Investigation:

As mentioned in the Summary of Theory, the CC amplifier is frequently used after a voltage amplifier and may derive its bias from the voltage amplifier. The circuit shown in Figure 9-5 uses the CE amplifier from Experiment 8 to drive the CC amplifier. Notice that the CC amplifier is dc coupled so it does not require bias resistors. Construct the circuit and measure the dc and ac parameters. Summarize your findings in a short report.

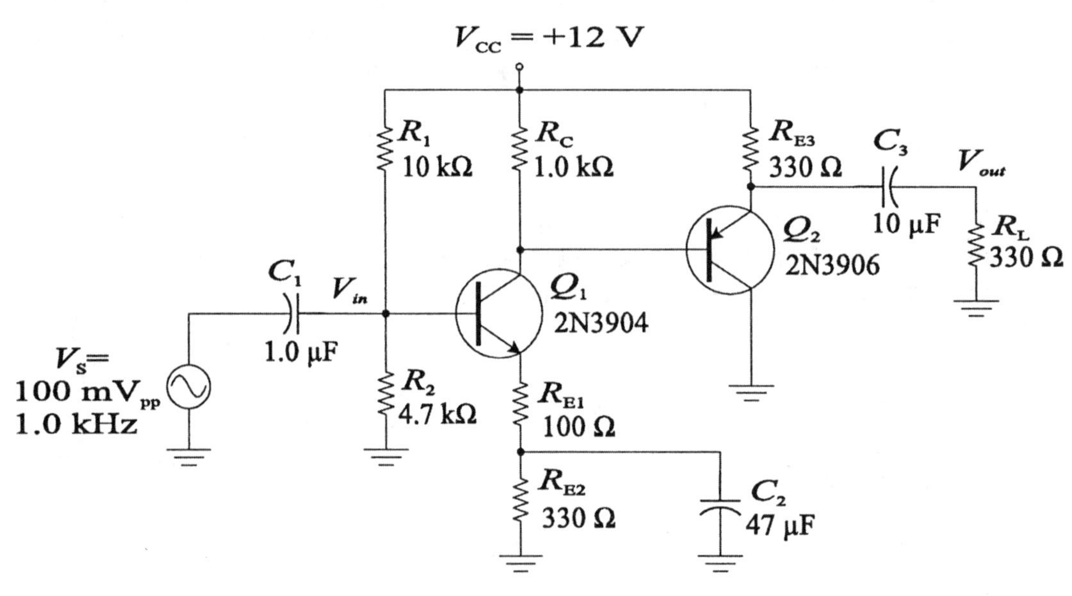

Figure 9-5

Name ______________
Date ______________
Class ______________

10 Transistor Switches

Reading:
Floyd and Buchla, *Analog Fundamentals: A Systems Approach*, Section 3-7

Objectives:
After performing this experiment, you will be able to:
1. Construct and test a basic one transistor circuit for its switching characteristics.
2. Explain how measurements can determine if a transistor is in cutoff or is saturated.
3. Add a second transistor to the circuit in objective 1 and measure the switching threshold.
4. Add hysteresis to the circuit and test both switching thresholds.

Summary of Theory:
An important application of transistors is in *switching circuits* used in digital systems and applications. The first large-scale use of digital circuits was in telephone systems. Today, computers form the most important application of switching circuits. Although nearly all of these circuits are digital integrated circuits, discrete transistor switching circuits are used when it is necessary to supply higher current or different voltages than can be furnished directly from the digital IC. Because of this versatility, discrete transistors are used to interface between different types of logic and various loads used in logic circuits. In addition to using low current or voltage to switch loads, transistor switches have other advantages over mechanical switches. They are more reliable than mechanical switches, cost less, provide faster switching times, and can provide isolation when a load is in a dangerous or remote location.

When transistors are used in switching applications, they are usually operated in either cutoff or saturation. *Cutoff* refers to the condition where the transistor acts as an open switch; *saturation* occurs when the transistor acts as a closed switch. If a transistor is in cutoff, there is no current in the collector circuit; if it is saturated, it is conducting as much as possible.

In this experiment, you will construct and test a one transistor switching circuit, then make improvements. The improvements involve: (1) avoiding the gradual switching of the one transistor circuit by adding a second transistor, (2) raising the voltage threshold where switching occurs, and (3) adding hysteresis. The switching threshold refers to the input voltage where the output changes from one state to another. In a switching circuit, hysteresis refers to two different thresholds, depending on whether the output is already saturated or already in cutoff. The advantage of hysteresis is that the switching is less susceptible to noise.

Materials Needed:
Resistors: one 330 Ω, one 1.0 kΩ, two 10 kΩ
One 10 kΩ potentiometer
Two small signal *npn* transistors (2N3904 or equivalent)
One LED

Procedure:

1. Measure and record the values of the resistors listed in Table 10-1. R_1 is a 10 kΩ potentiometer and is not listed in the table. R_{C1} is used in step 4.

Table 10-1

Resistor	Listed Value	Measured Value
R_B	10 kΩ	
R_C	1.0 kΩ	
R_{C1}	10 kΩ	
R_E	330 Ω	

2. Ideally, a transistor switching circuit should operate either in cutoff or in saturation. The circuit shown in Figure 10-1 is a basic transistor amplifier. It can easily be set for operation at these extremes by varying the potentiometer (R_1). However, it can also operate *between* cutoff and saturation, a condition not desirable in a switching circuit. Compute the collector-emitter voltage (V_{CE}) at cutoff and saturation and the voltage across the collector resistor at saturation (V_{RC}). Compute I_{sat} by assuming a 2.0 V drop across the LED and 0.1 V across the transistor. Show the computed values in Table 10-2.

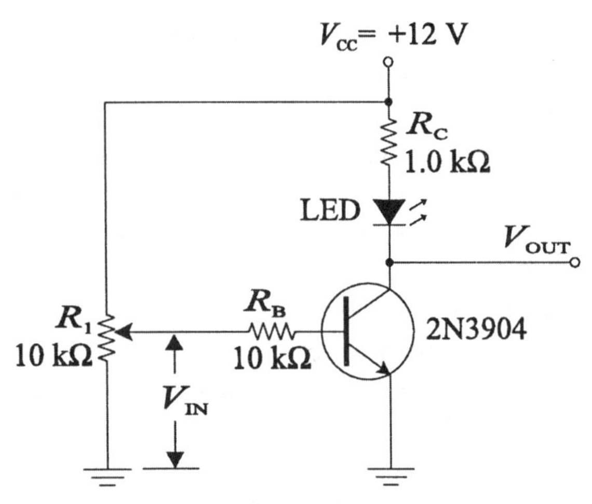

Figure 10-1

Table 10-2

Quantity	Computed Value	Measured Value
$V_{CE(cutoff)}$		
$V_{CE(sat)}$		
$V_{RC(sat)}$		
I_{sat}		

3. Construct the circuit in Figure 10-1 and observe the effect of varying the potentiometer. Set the potentiometer to the minimum value and measure V_{CE} at cutoff (the LED should be off). Then set the potentiometer for maximum (LED on), and measure V_{CE} (saturation) and the voltage across the 1.0 kΩ collector resistor, V_{RC} (saturation). The transistor is saturated since it can supply no more current in the collector circuit. Record measurements in Table 10-2 and compare to the computed values from step 2.

4. In this step, you will add a second transistor, causing the switching action to improve dramatically. The circuit is shown in Figure 10-2. Notice that the 1.0 kΩ resistor is now the collector resistor for Q_2. The circuit works as follows. When V_{IN} is very low, Q_1 is off since it does not have sufficient base current. Q_2 will be in saturation because it can obtain ample base current through R_{C1}, so the LED is on. As the base voltage for Q_1 is increased, Q_1 begins to conduct. As Q_1 approaches saturation, the base voltage of Q_2 drops, causing it to go from a saturated to cutoff condition rapidly. The output voltage of Q_2 drops and the LED goes out. Construct the circuit, and test it as described in the next step.

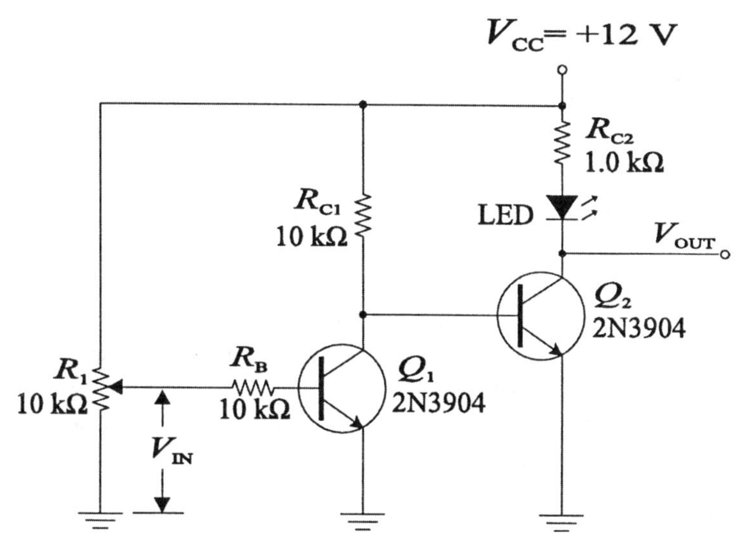

Figure 10-2

Table 10-3

Quantity	Measured Value
V_{IN} (LED on)	
V_{OUT} (LED on)	
V_{IN} (threshold)	
V_{OUT} (threshold)	

5. Set the potentiometer so that V_{IN} is a minimum (0 V). Since Q_1 is *off*, the LED should be be *on*. Measure V_{IN} and V_{OUT} and record the readings in Table 10-3 in the first two rows. Monitor V_{IN} and *slowly* increase V_{IN} while watching the LED. You should observe that there is no dim condition for the LED - it will suddenly go off as the input voltage is increased. Record V_{IN} and V_{OUT} at the threshold where the LED just turns off. Notice that the output voltage indicates Q_2 is either in saturation or in cutoff.

6. In step 5, the switching threshold was distinct; however the threshold voltage is rather low and it is susceptible to noise on the input. There is another simple improvement you can make to this circuit. The improvement is shown in Figure 10-3. The common-emitter resistor, R_E, will raise the threshold voltage. In addition, because of the different saturation currents of the two transistors, the threshold will be different when the output is in cutoff than when the output is saturated. This is a very useful feature, called hysteresis, and is characteristic of Schmitt trigger circuits (to be investigated in Experiment 22). Modify the circuit from step 5 by adding the 330 Ω resistor; then test it according to the procedure in the next step.

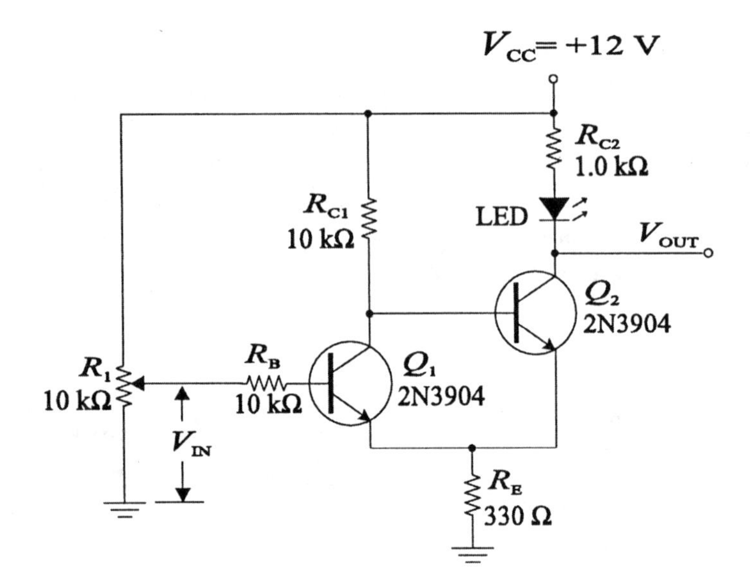

Figure 10-3

Table 10-4

Quantity	Measured Value
V_{IN} (LED on)	
V_{OUT} (LED on)	
V_{IN} (upper threshold)	
V_{OUT} (upper threshold)	
V_{IN} (lower threshold)	
V_{OUT} (lower threshold)	

7. Set the potentiometer so that V_{IN} is a minimum (0 V). The LED should be *on*. Measure V_{IN} and V_{OUT} and record the readings in Table 10-4. The reading of V_{OUT} is higher than in the previous circuit but the transistor is still saturated (why?). Test the upper threshold by monitoring V_{IN} as you *slowly* increase V_{IN}. You will see the LED go out suddenly. Record this as the upper threshold voltage. Measure and record V_{OUT} and confirm that the transistor is in cutoff. Now *slowly* reduce V_{IN}. Notice that the LED stays out until the voltage is much lower. When it comes on, record V_{IN} and V_{OUT} at the lower threshold.

Conclusion:

Evaluation and Review Questions:

1. Give at least three advantages of transistor switching circuits.

2. What is the purpose of R_B in Figure 10-1?

3. Assume you wanted to determine the base current in Figure 10-1. What voltage measurement would you make to do this indirectly?

4. In step 6, it was stated that the saturation current for the two transistors is different. Why?

5. Assume a student measured the collector current in a saturated transistor as 10 mA and the base current as 0.25 mA. Why can't these measurements be used to determine the β_{DC} of the transistor?

For Further Investigation:

A transfer curve is a plot of the output of a circuit (plotted along the *y*-axis) versus the input (plotted along the *x*-axis). Set up a data table for the circuit in Figure 10-3 and record the input and output voltage as the input is increased and decreased. Take sufficient data that you know precisely what the output voltage is for the range of input voltages. Then plot the transfer curve for the circuit in Plot 10-1. Label the axes of your plot.

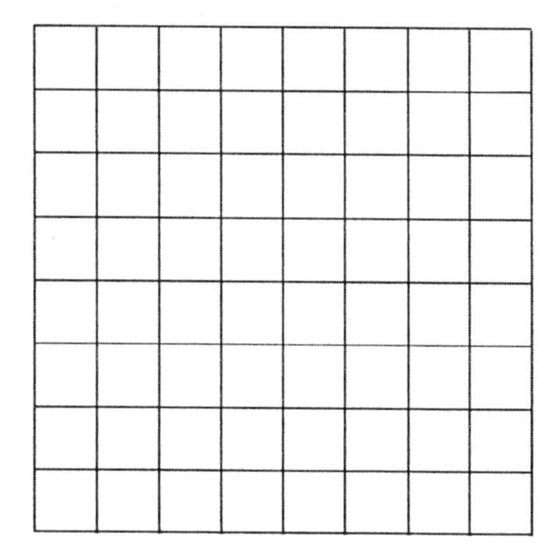

Plot 10-1

Name ______________
Date ______________
Class ______________

11 JFET Characteristics

Reading:
Floyd and Buchla, *Analog Fundamentals: A Systems Approach*, Sections 4-1 and 4-2

Objectives:
After performing this experiment, you will be able to:
1. Measure and graph the drain characteristic curves for a junction field-effect transistor (JFET).
2. Measure $V_{GS(off)}$ and I_{DSS} for a JFET.
3. Connect a JFET as a two-terminal constant-current source to maintain constant illumination in an LED.

Summary of Theory:
The bipolar junction transistor (BJT) uses base current to control collector current. Unlike the BJT, the field-effect transistor (FET) is a voltage-controlled device that uses an electrostatic field to control current. The FET begins with a doped piece of silicon called a *channel.* On one end of the channel is a terminal called the *source* and on the other end of the channel is a terminal called the *drain.* Current in the channel is controlled by a voltage applied to a third terminal called the *gate.* Field-effect transistors are classified as either junction-gate (JFET) or insulated-gate (IGFET) devices. The JFET has a reverse-biased diode at the gate whereas the IGFET uses a thin glass insulting layer. Since the gate circuit of either type of FET draws almost no current, the input resistance is extremely high. Both types have similar ac characteristics but differ in biasing methods.

The gate of a JFET is made of the opposite type of material than the channel, forming a *pn* junction between the gate and channel. Application of a reverse-bias on this junction decreases the conductivity of the channel, reducing the source-drain current. The gate diode should never be forward-biased. The JFET comes in two forms, *n*-channel and *p*-channel. The *n*-channel is distinguished on drawings by an inward arrow on the gate connection while the *p*-channel has an outward pointing arrow on the gate as shown in Figure 11-1.

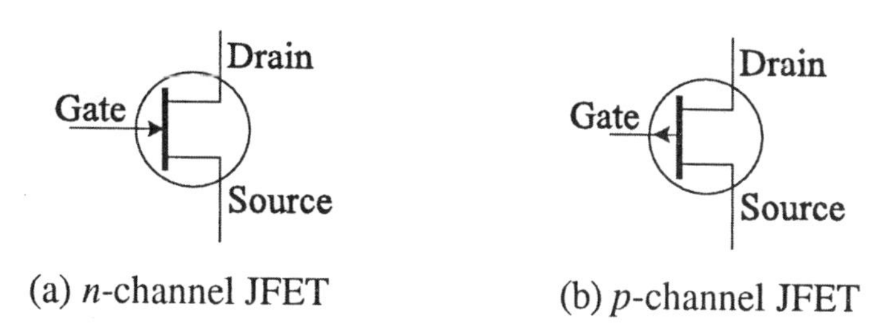

Figure 11-1

The characteristic drain curves for a JFET exhibit several important differences from the BJT. In addition to the fact that the JFET is a voltage-controlled device, the JFET is a normally ON device. In other words, a reverse-bias voltage must be applied to the gate-source *pn* junction in order to close off the channel and prevent drain-source current. When the gate is shorted to the source, there is maximum allowable drain-source current. This current is called I_{DSS} for Drain-Source current with gate Shorted. Another important difference is that the JFET exhibits a region on its characteristic curve where drain current is proportional to the drain-source voltage. This region, called the ohmic region, has important applications as a voltage-controlled resistance. (An example of the application of the ohmic region to automatic gain control will be tested in Experiment 15 – For Further Investigation.)

A useful specification for estimating the gain of a JFET is called the *transconductance*, which is abbreviated g_m. Recall that conductance is the reciprocal of resistance. Since the *output current* is controlled by an *input voltage*, it is useful to think of any FET as a transconductance amplifier. The transconductance can be found by dividing a small change in the output current by a small change in the input voltage. That is:

$$g_m = \frac{\Delta I_D}{\Delta V_{GS}}$$

You will use the data from this experiment in Experiment 12, so you should save your JFET.

Materials Needed:

One 100 Ω resistor
One 10 kΩ resistor
One 2N5458 *n* channel JFET (or equivalent). (Save the JFET for Experiment 12.)
One LED
One milliammeter 0-10 mA range

Procedure:

1. Measure and record the value of the resistors listed in Table 11-1. R_1 is used for protection in case the JFET is forward-biased accidently. R_2 serves as a current-sensing resistor.

2. Construct the circuit shown in Figure 11-2. Start with V_{GG} and V_{DD} at 0 V. Connect a voltmeter between the drain and source. Keep V_{GG} at 0 V and slowly increase V_{DD} until V_{DS} is 1.0 V. (V_{DS} is the voltage between the transistor's drain and source.)

Table 11-1

Resistor	Listed Value	Measured Value
R_1	10 kΩ	
R_2	100 Ω	

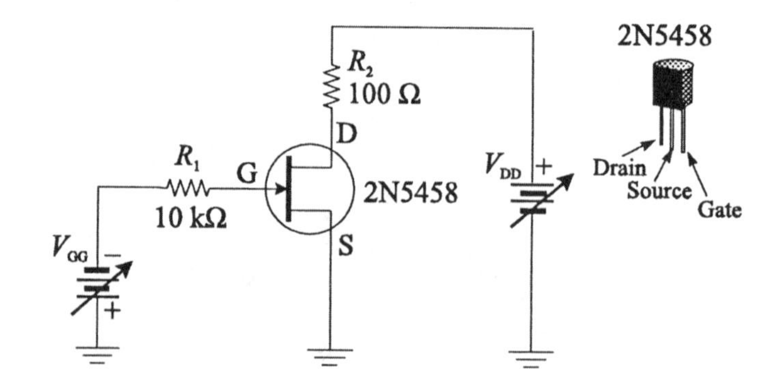

Figure 11-2

3. With V_{DS} at 1.0 V, measure the voltage across R_2 (V_{R2}). Compute the drain current, I_D, by applying Ohm's law to R_2. Note that the current in R_2 is the same as I_D for the transistor. Use the measured voltage, V_{R2}, and the measured resistance, R_2, to determine I_D. Enter the measured value of V_{R2} and the computed I_D in Table 11-2 under the columns labeled Gate Voltage = 0 V.

Table 11-2

	Gate Voltage = 0 V		Gate Voltage = – 0.5 V		Gate Voltage = – 1.0 V		Gate Voltage = – 1.5 V	
V_{DS}	V_{R2}	I_D	V_{R2}	I_D	V_{R2}	I_D	V_{R2}	I_D
1.0 V								
2.0 V								
3.0 V								
4.0 V								
6.0 V								
8.0 V								

4. Without disturbing the setting of V_{GG}, slowly increase V_{DD} until V_{DS} is 2.0 V. Then measure and record V_{R2} for this setting. Compute I_D as before and enter the measured voltage and computed current in Table 11-2 under the columns labeled Gate Voltage = 0 V.

5. Repeat step 4 for each value of V_{DS} listed in Table 11-2.

6. Adjust V_{GG} for –0.5 V. This applies –0.5 V between the gate and source because there is almost no gate current into the JFET and almost no voltage drop across R_1. Reset V_{DD} until V_{DS} = 1.0 V. Measure V_{R2} and compute I_D as before. Enter the values in Table 11-2 under the columns labeled Gate Voltage = –0.5 V.

7. Without changing the setting of V_{GG}, adjust V_{DD} for each value of V_{DS} listed in Table 11-2 as before. Compute the drain current at each setting and enter the voltage and current values in Table 11-2 under the columns labeled Gate Voltage = –0.5 V.

8. Adjust V_{GG} for –1.0 V. Reset V_{DD} until V_{DS} = 1.0 V. Repeat step 7, entering the data in the columns labeled Gate Voltage = –1.0 V.

9. Adjust V_{GG} for –1.5 V. Reset V_{DD} until V_{DS} = 1.0 V. Repeat step 7, entering the data in the columns labeled Gate Voltage = –1.5 V.

10. The data in Table 11-2 represent four drain characteristic curves for your JFET. The drain characteristic curve is a graph of V_{DS} versus I_D for a constant gate voltage. Plot the four drain characteristic curves on Plot 11-1. Choose a scale for I_D that allows the largest current observed to fit on the graph. Label each curve with the gate voltage it represents.

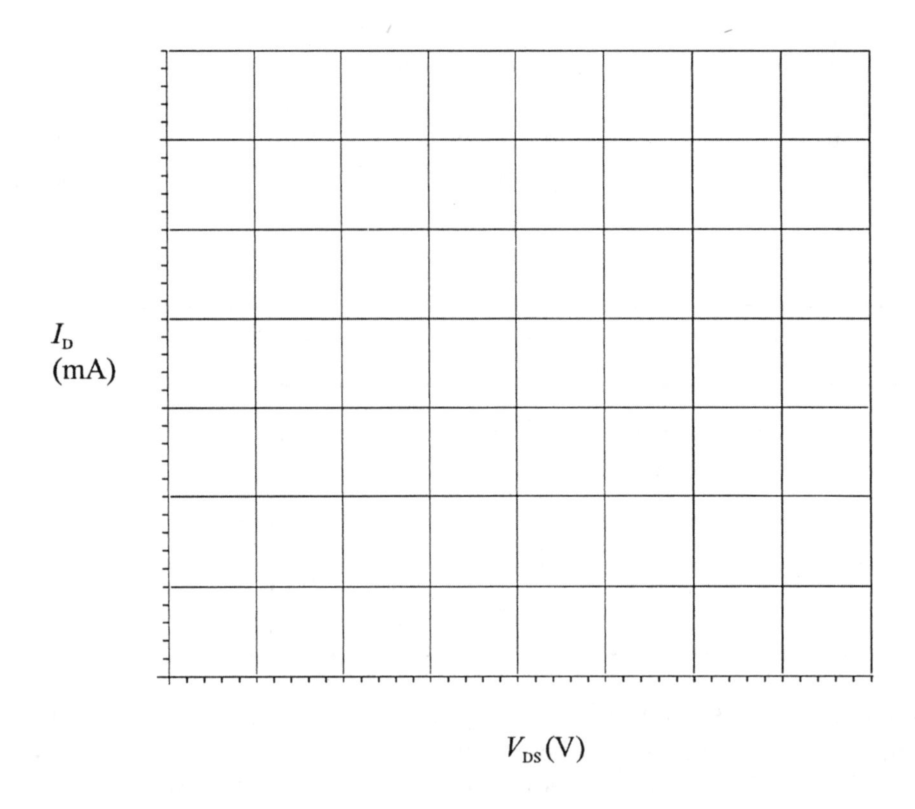

Plot 11-1

11. In this step you will determine the value of $V_{GS(off)}$. Set V_{DD} for +12 V and V_{GG} for 0 V. Monitor the voltage across R_2 and slowly increase the negative gate voltage. When the voltage across R_2 reaches zero, note the gate voltage. Record this value in Table 11-3 as $V_{GS(off)}$. Record I_{DSS} from reading Plot 11-1. These are the key parameters for your JFET that will be used in Experiment 12.

Table 11-3

Measured JFET Parameters
$V_{GS(off)}$ =
I_{DSS} =

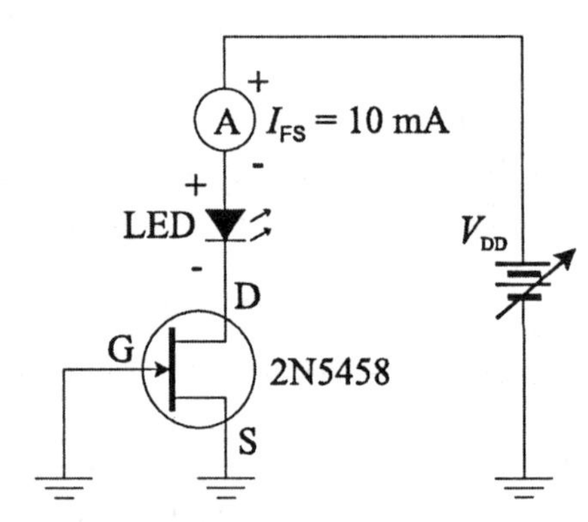

Figure 11-3

12. In this step, you can observe a JFET connected as a two-terminal constant current source. Construct the circuit shown in Figure 11-3. Monitor the drain voltage while you increase the drain power supply from 0 V to +15 V. Notice the drain voltage where constant current begins (this point is imprecise but about the same as the absolute value of $V_{GS(off)}$). Compare the ammeter reading with the maximum current found in step 3. Observations:

____

Conclusion:

Evaluation and Review Questions:

1. (a) Explain how to find I_{DSS} from the characteristic curves of a JFET.

   (b) What was the maximum current in the LED in step 12?

2. (a) Does the experimental data indicate that the transconductance is a constant at all points?

   (b) From your experimental data, what evidence indicates that a JFET is a nonlinear device?

3. Look up the meaning of pinch-off voltage when $V_{GS} = 0$. Note that the magnitude of V_{GS} is equal to the magnitude of V_p so we can use the characteristic curve for $V_{GS} = 0$ to determine V_p. Using the data from this experiment, determine the pinch-off voltage for your JFET.

4. Why should a JFET be operated with only reverse bias on the gate source?

5. Compare the characteristic curve for a bipolar transistor (Experiment 6) with the characteristic curve observed in this experiment for a JFET.

For Further Investigation:

Using the test circuit shown in Figure 11-2, test the effect of varying V_{GS} with V_{DD} held at a constant +12 V. Tabulate a set of data of I_D as a function of V_{GS} (Table 11-4). Start with $V_{GS} = 0.0$ V and take data every – 0.5 V until there is no appreciable drain current. Then graph the data on Plot 11-2. This curve is the *transconductance* curve for your JFET. The result is nonlinear because the gate-source voltage is proportional to the square root of the drain current. To illustrate this, compute the square root of I_D and record in Table 11-4. Plot the square root of the drain current, ($\sqrt{I_D}$), as a function of the gate-source voltage on Plot 11-3.

Table 11-4

V_{GS} (measured)	I_D (measured)	$\sqrt{I_D}$ (computed)
0 V		
– 0.5 V		
– 1.0 V		
– 1.5 V		
– 2.0 V		
– 2.5 V		
– 3.0 V		
– 3.5 V		
– 4.0 V		
– 4.5 V		
– 5.0 V		

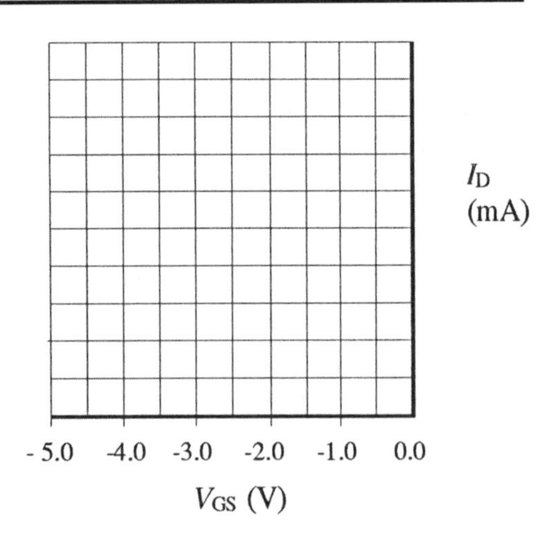

Plot 11-2

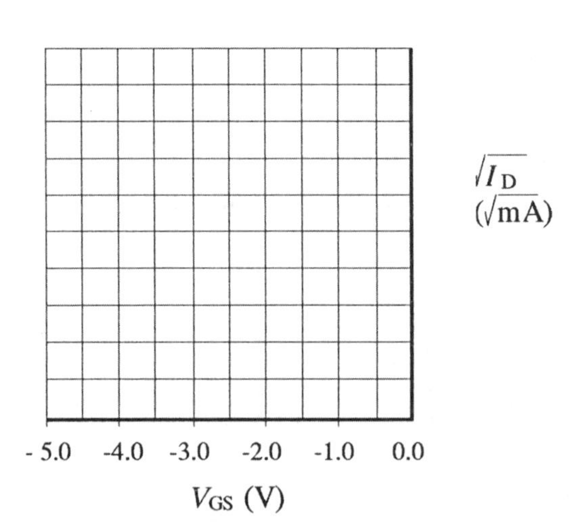

Plot 11-3

Name ______________
Date ______________
Class ______________

12 JFET Biasing

Reading:
Floyd and Buchla, *Analog Fundamentals: A Systems Approach*, Section 4-3

Objectives:
After performing this experiment, you will be able to:

1. Determine $V_{GS(off)}$ indirectly from a self-bias circuit.
2. Specify the source resistor for a self-biased JFET that will provide a specified I_D and draw the load line for the resistor.
3. Specify the source resistor for a JFET using voltage-divider bias and test the circuit.

Summary of Theory:
As you know, *bias* is the application of dc voltages to set up the proper quiescent conditions for circuit operation. A satisfactory bias circuit for a FET depends on its type. With depletion-mode devices, which include all JFETs and some D-MOSFETs, the gate must be reverse biased (or zero biased) with respect to the source. These devices are normally *on* – they are turned *off* by applying reverse bias to the gate. Most MOSFETs operate as enhancement mode devices (all E-MOSFETs and some D-MOSFETs) and require bias to turn them *on*.

Self-bias is the most common type of bias for JFETs and is illustrated in Figure 12-1(a). The drain current, I_D, is in the source resistor, creating a voltage $V_S = I_D R_S$ at the source terminal.

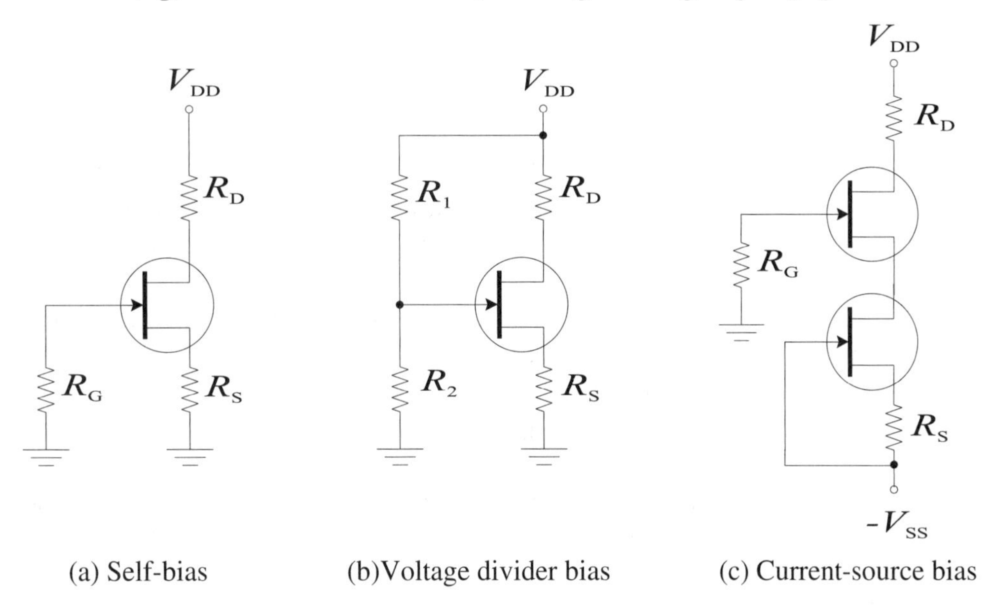

(a) Self-bias (b)Voltage divider bias (c) Current-source bias

Figure 12-1

Since the gate is at ground potential (0 V), the gate-source voltage must have the same magnitude but opposite sign to the voltage drop across R_S (by Kirchhoff's voltage law). For an *n*-channel device, $V_{GS} = -V_S$. This provides the required reverse bias on the gate. Self-bias tends to compensate for different device characteristics between various FETs. For example, if a device with higher transconductance is put in the circuit, the drain current increases along with the voltage drop across R_S. This increased voltage tends to bias the FET off, compensating for the higher transconductance.

An even more stable form of bias combines self-bias with voltage-divider bias as illustrated in Figure 12-1(b). The voltage-divider connected to the gate biases the gate at some positive voltage. Unlike bipolar transistors, the JFET draws almost no input current, so the divider resistors can be much larger. The source voltage must still be more positive than the gate in order to establish the proper gate-source reverse bias. To accomplish this, the source resistor is made large enough to develop a positive voltage with respect to the gate (much larger than in self-bias). The net result is that transistor variations have less effect on the operating point than self-bias, producing a more stable form of biasing. The drain current is fairly independent of the transistor; however, the drain-source voltage must be large enough to assure that the transistor is operating in the constant-current region.

Another very stable form of biasing is called *current-source biasing*, illustrated in Figure 12-1(c). In this form of bias, the FET is connected in *cascode* with a current source; the lower transistor provides constant current for the upper FET. Example 4-8 in the text illustrates current-source biasing using an *npn* biasing transistor. It is also discussed in the For Further Investigation section of this experiment and in the For Further Investigation section of Experiment 14.

Materials Needed:

Resistors: one 2.2 kΩ, one 330 kΩ, one 1.0 MΩ, two values to be determined by student
One 1.0 kΩ potentiometer
One 2N5458 JFET
One milliammeter 0 - 10 mA

For Further Investigation:

one additional 2N5458 JFET

Procedure:

1. In this step, you will measure the values of $V_{GS(off)}$ and I_{DSS} in a slightly modified circuit from Experiment 11 (and introduce self-bias at the same time!). You will determine the value of $V_{GS(off)}$ indirectly – by measuring two points on the transfer characteristic and computing $V_{GS(off)}$ from the idealized model for FET behavior. If you have saved the FET from Experiment 11, and know the values of $V_{GS(off)}$ and I_{DSS}, you may want to compare those measurements with the results from this step.
   (a) Set up the circuit shown in Figure 12-2. Set the potentiometer for zero ohms and record the current reading on the milliammeter. This is I_{DSS}. (Why?)

   (b) Increase the resistance until the milliammeter reads 50% of I_{DSS}. At this point, measure V_{GS}. Multiply the reading of V_{GS} by 3.4; this is $V_{GS(off)}$.[1]

---

[1] As an interesting aside, a FET appears to have an internal resistance, analogous to r_e' of a bipolar transistor. When R is adjusted for one-half I_{DSS}, the transconductance of the FET is equal to the reciprocal of the resistance R.

Enter the values of $V_{GS(off)}$ and I_{DSS} into Table 12-1.

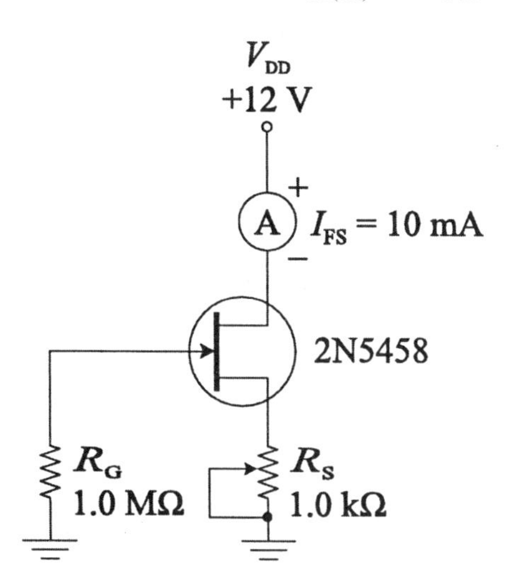

Figure 12-2

Table 12-1

Measured JFET Parameters
$V_{GS(off)}$ =
I_{DSS} =

2. Plot 12-1 shows a normalized graph of I_D verses V_{GS}. Normalized means that the values on the curve have been set to dimensionless numbers that can be converted to actual measured quantities by a scaling operation. In the normalized curve shown, $V_{GS(off)}$ and I_{DSS} are both set to the value 1.0. Convert the scale to a practical curve for your JFET by multiplying the values shown on the x-axis by $V_{GS(off)}$ and the values shown on the y-axis by I_{DSS}. Place the practical values on the graph in the space provided.

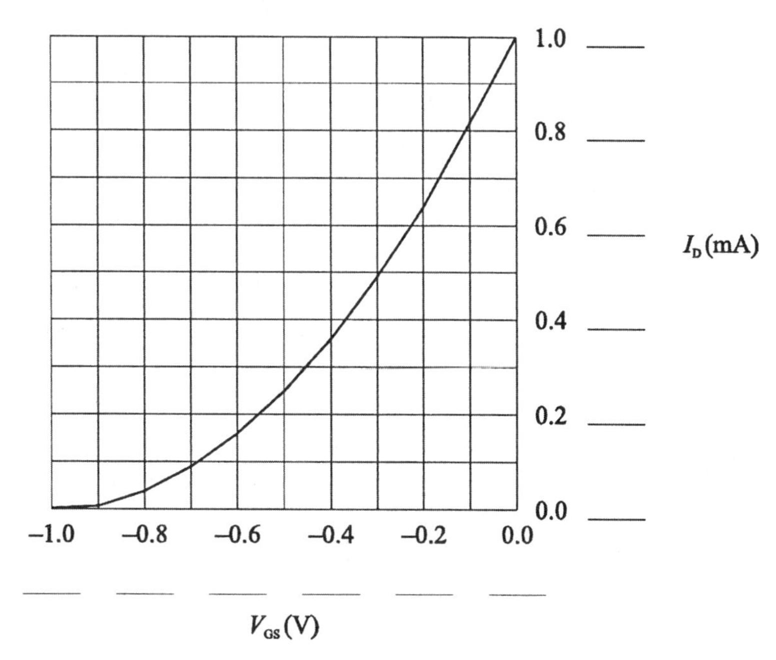

Plot 12-1

3. Besides midpoint bias, another reasonable choice for a source resistor for self-bias is to select one that has the same resistance as the drain-source channel resistance of the ideal JFET in the ohmic region. This resistance is given by the equation

$$R_S = \frac{|V_{GS(off)}|}{I_{DSS}}$$

where $|V_{GS(off)}|$ means the absolute (unsigned) value of $V_{GS(off)}$. The resistor given by this equation produces a drain current of approximately 38% of I_{DSS}. From your data, determine the value of the nearest standard source resistance that meets this criterion.

R_S = _______________ (nearest standard resistance)

Draw the load line on Plot 12-1 for the resistor determined above. Indicate the Q-point on the plot (see text, Figure 4-13). From the plot, predict I_D for the circuit.

I_D = _______________ (predicted from plot)

4. Replace the potentiometer shown in Figure 12-2 with the source resistor determined above. Measure the drain current. It should be in good agreement with the value predicted in step 3.

I_D = _______________ (measured)

5. Voltage-divider bias uses a combination of a voltage divider on the gate plus a source resistor to establish the source at a higher potential than the gate. Assume you wish to maintain the same Q-point as in step 3 (the self-bias case) but with voltage-divider bias. The divider is shown in Figure 12-3. Calculate a source resistor that will keep the same gate-source voltage as before. The steps are:

(a) Compute the voltage on the gate from the voltage-divider rule. Enter in Table 12-2.

(b) Add this gate voltage to the absolute value of V_{GS} as determined in step 3 for the Q-point. This represents the desired source voltage. Enter this as the computed value in Table 12-2.

(c) The drain current should be the same as before. Using this value, apply Ohm's law to find the value of the source resistor. Enter the computed resistance in Table 12-2.

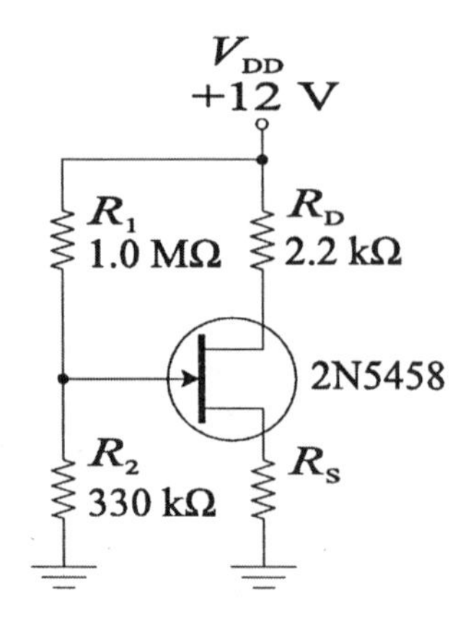

Figure 12-3

Table 12-2

Voltage-divider biased JFET	Computed Value	Measured Value
V_G		
V_S		
R_S		
I_D		

6. Construct the circuit using the value of source resistor you found in step 5. Measure and record the parameters listed in Table 12-2. You will need to make sure your voltmeter has at least 10 MΩ input resistance or it will load the circuit and produce invalid readings. The drain current is measured *indirectly* by dividing the measured source voltage by the source resistor (drain and source current are the same). You should find that the gate-source voltage and the drain current are nearly the same as in the self-biased case.

Conclusion:

Evaluation and Review Questions:

1. In step 1, you indirectly determined the value of $V_{GS(off)}$ by using one-half I_{DSS} in a self-biased circuit. Why isn't it practical to measure $V_{GS(off)}$ by increasing R until the transistor ceases to conduct?

2. Contrast the biasing of a FET with biasing a bipolar transistor. Why can't the same bias circuits be used for both types of transistors?

3. (a) For the self-bias circuit shown in Figure 12-2, what is the purpose of R_G?

   (b) Will the circuit work with it replaced with a short?

4. What is the significant difference between the voltage dividers used in JFET biasing than those used for BJTs?

5. How does the biasing for a JFET differ from the biasing for an E-MOSFET?

For Further Investigation:

The most consistent bias is formed by adding a current-source to the circuit. This is particularly useful for a source-follower circuit (analogous to the emitter-follower) because the current source has extremely high resistance and doesn't load the source. (This circuit will be investigated in Experiment 13). The current-source bias circuit shown in Figure 12-4 uses the lower transistor as the current source. Set the current to the same value used in the experiment by using the source resistor from step 3 for R_S. Measure the voltages for the circuit and determine if the Q-point is the same as before.

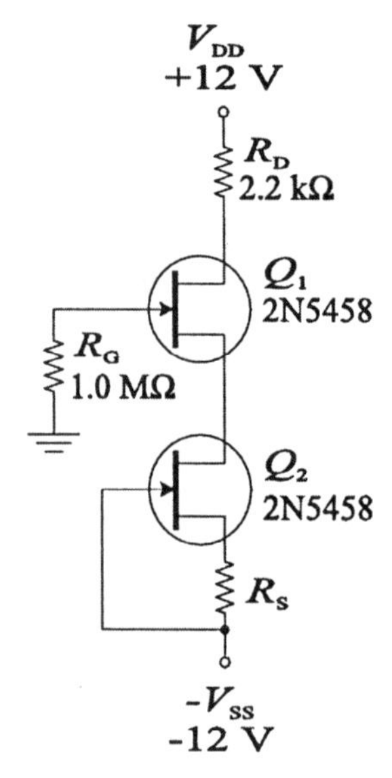

Figure 12-4

Name ______________
Date ______________
Class ______________

13 FET Amplifiers

Reading:
Floyd and Buchla, *Analog Fundamentals: A Systems Approach*, Section 4-6

Objectives:
After performing this experiment, you will be able to:
1. Measure dc and ac parameters for self-biased common-source and common-drain amplifiers.
2. Test a common-drain amplifier with current-source biasing.

Summary of Theory:
Field-effect transistors are available as either JFETs or MOSFETs. In Experiment 12, you tested bias methods for JFETs. MOSFETs are biased according to type; D-MOSFETs frequently have no bias (but can be biased) and E-MOSFETs are biased by the same techniques used for bipolar transistors.

FETs are used in common-source, common-gate, and common-drain (source-follower) configurations similar to the bipolar common-emitter, common-base, and common collector (emitter-follower) amplifiers. The major advantage of FETs over bipolar transistors is their extremely high input resistance. In this experiment, you will test ac configurations of FET amplifiers, beginning with a self-biased common-source amplifier as illustrated in Figure 13-1(a), then a self-biased common-drain amplifier as illustrated in Figure 13-1(b). You will then add a current-source load and observe the improvement in follower action. Finally, a D-MOSFET amplifier will be tested in the For Further Investigation.

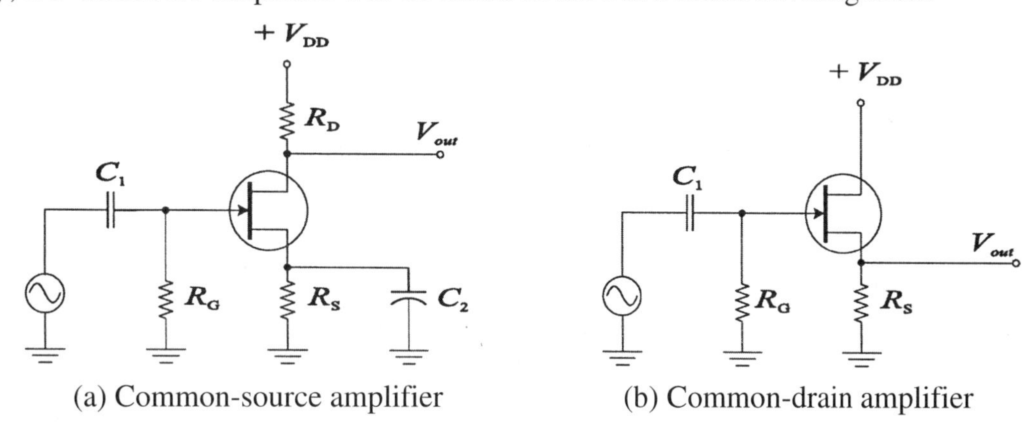

(a) Common-source amplifier (b) Common-drain amplifier

Figure 13-1

Materials Needed:
Resistors: two 1.0 kΩ, one 3.3 kΩ, one 10 kΩ, one 100 kΩ, one 1.0 MΩ, one to be determined by student
Two 2N5458 *n*-channel JFETs (or equivalent)
Capacitors: one 0.1 μF, one 1.0 μF, one 10 μF

For Further Investigation:
One 1 kΩ potentiometer
One 620 Ω resistor

Procedure:
Common-Source JFET Amplifier

1. Measure and record the resistance of the resistors listed in Table 13-1.

2. Construct the common-source amplifier shown in Figure 13-2. Set the signal generator for a 500 mV_{pp} sine wave at 1.0 kHz. Check the amplitude and frequency with your oscilloscope.

Table 13-1

Resistor	Listed Value	Measured Value
R_S	1.0 kΩ	
R_D	3.3 kΩ	
R_G	1.0 MΩ	
R_L	10 kΩ	

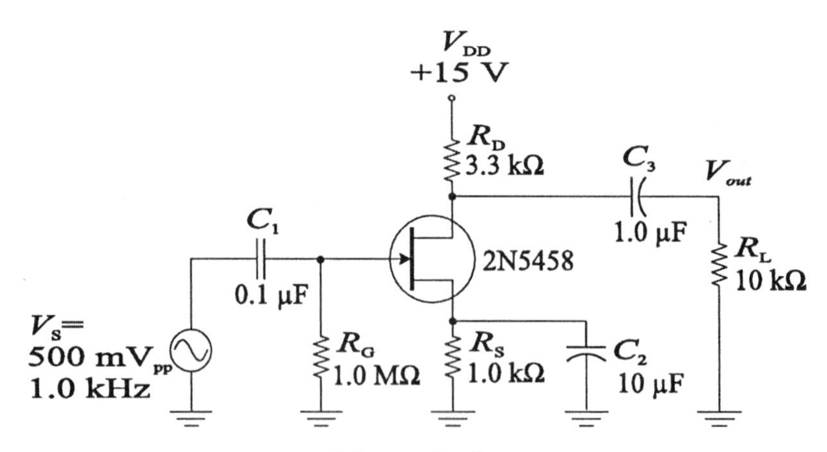

Figure 13-2

3. Measure the dc voltage at the drain, source, and gate. Use the source voltage and source resistance to compute I_D. Enter the data in Table 13-2. Compare the input and output ac voltage by viewing V_{in} and V_{out} simultaneously on a two-channel oscilloscope. Measure the voltage gain and note the phase difference (0° or 180°) between the input and output signal.

Table 13-2

Data for Common-Source Amplifier	DC values	AC values
Gate voltage, V_G		
Source voltage, V_S		
Drain voltage, V_D		
Drain current, I_D		
Input voltage, V_{in}		
Output voltage, V_{out}		
Voltage gain, A_v		
Phase difference		

4. Change the source resistor from 1.0 kΩ to the value you found in Experiment 12, step 3, for the self-bias case. If you do not have the same FET or did not record the resistor, use a 620 Ω resistor in place of R_S. You should observe a slight increase in gain with the smaller resistor. Can you explain this gain increase? (*Hint:* consider g_m.)

____

5. Now change the load resistor from 10 kΩ to 100 kΩ. Explain the observed gain change.

____

Common-Drain Self-Biased JFET Amplifier

6. Change the circuit to the self-biased common-drain configuration shown in Figure 13-3. The drain is connected directly to +15 V. Measure the dc voltage at the drain, source, and gate and compute I_D. Observe the input and output ac voltage with the oscilloscope. Measure the voltage gain and note the phase. Enter the data in Table 13-3.

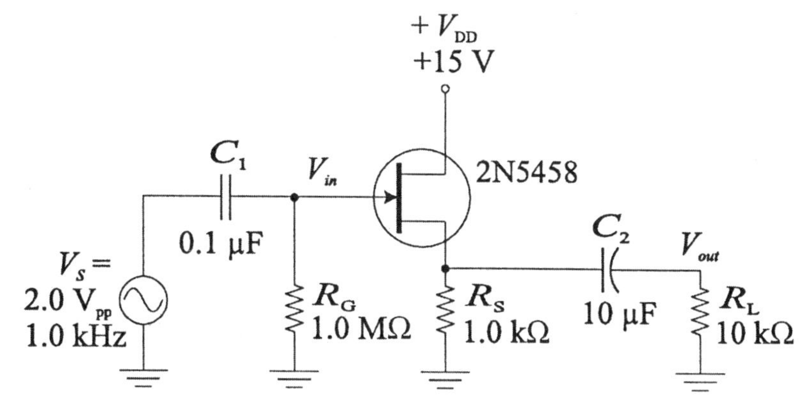

Figure 13-3

Table 13-3

Data for Common-Drain Self-biased Amplifier	DC values	AC values
Gate voltage, V_G		
Source voltage, V_S		
Drain voltage, V_D		
Drain current, I_D		
Input voltage, V_{in}		
Output voltage, V_{out}		
Voltage gain, A_v		
Phase difference		

Common-Drain Current-Source JFET amplifier

7. As you observed in step 6, the voltage gain is much less than 1.0 due to the transconductance, g_m. The reciprocal of g_m ($1/g_m$) is analogous to r_e' of a bipolar transistor, but with higher resistance. This forms a voltage divider with R_S as shown in Figure 13-4. To improve the gain, the divider is constructed with a very much larger resistance in the form of a JFET current source instead of a fixed resistor as shown in Figure 13-5. Modify the self-biased circuit to include the current source. Remove the output coupling capacitor and the load resistor and lift the ground end of R_S. Connect the drain of a JFET current source to this point as shown. Notice that the output is taken at the bottom of R_{S1}.

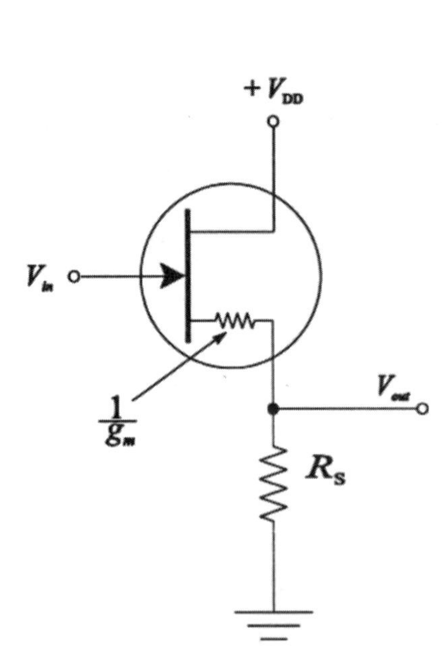

Figure 13-4 $1/g_m$ and R_S form a voltage divider.

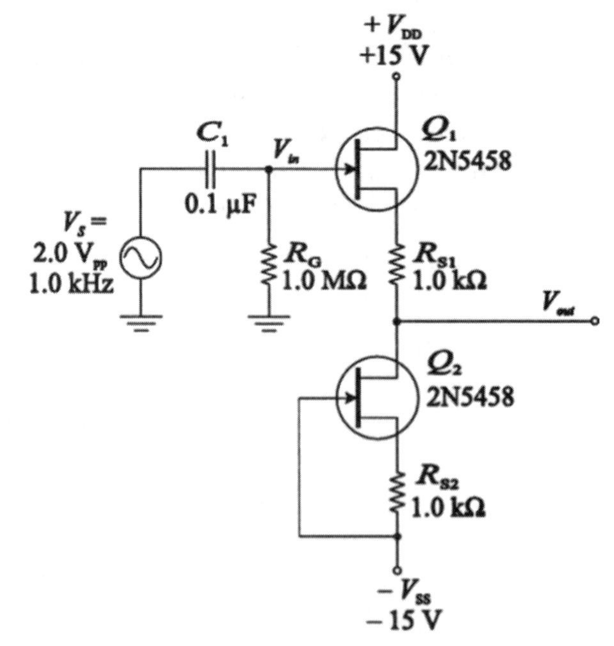

Figure 13-5 Q_2 is a current source that has very high drain-source resistance.

8. Measure and record the parameters listed in Table 13-4 for the common-drain, current-source amplifier. Note the change in the gain.

Table 13-4

Data for Common-Drain Current-Source Amplifier	DC values	AC values
Q_1 Gate voltage, V_G		
Q_1 Source voltage, V_S		
Q_1 Drain voltage, V_D		
Q_2 Gate voltage, V_G		
Q_2 Source voltage, V_S		
Q_2 Drain voltage, V_D		
Drain current, I_D		
Input voltage, V_{in}		
Output voltage, V_{out}		
Voltage gain, A_v		
Phase difference		

9. You should have found that the dc voltage at the drain of Q_2 (the output) is nearly zero. (Why?) With almost no offset, the load resistor can be directly connected to the output. Test the effect of the 10 kΩ load on the amplifier. Then test the signal clipping point by increasing the input signal from the generator. Summarize your observations.

Conclusion:

Evaluation and Review Questions:

1. (a) What advantage does a common-drain amplifier have compared to a common-emitter amplifier?

   (b) What disadvantage does a common-drain amplifier have compared to a common-emitter amplifier?

2. Compare the common-source and common drain amplifiers tested in this experiment. What are the significant differences between them? What things do they have in common?

3. For the common-source amplifier (Figure 13-2), what would you expect to happen to the dc and ac parameters if the bypass capacitor, C_2, were open?

4. Assume you want to modify the common-source amplifier in Figure 13-2 for current-source biasing.
   (a) Is the source resistor, R_S, required?

   (b) Is the bypass capacitor, C_2, required?

   (c) What advantage and disadvantage would result from the change?

5. For the common-drain current-source biasing circuit shown in Figure 13-5,
   (a) Why is it useful to have the source resistor, R_{S1}, equal to R_{S2}?

   (b) Why was the gain much better with current-source biasing than self-bias?

   (c) Estimate the input and output resistance of the amplifier.

For Further Investigation:
A simple modification of the circuit in Figure 13-5 can make a useful buffer for dc signals. Transducers frequently have a dc output and a high equivalent internal resistance. In this case, a dc amplifier set up as a buffer is quite useful. Modify the circuit in Figure 13-5 to accept a dc input by eliminating the input coupling capacitor. To simulate a high internal resistance, remove R_G from ground and connect it in series with a dc source that you can vary between −3 V to + 3 V (You may need to set up a potentiometer between the power supplies to do this.) Change R_{S2} to a 1 kΩ potentiometer (set up as a rheostat) in series with a 620 Ω resistor. If the transistors are reasonably well matched, you should be able to have the output track the input. Set up the circuit and test it over a range of input voltages from −3 V to + 3 V. Investigate this circuit and summarize your findings in a short report.

Name ______________
Date ______________
Class ______________

14 JFET Applications

Reading:
Floyd and Buchla, *Analog Fundamentals: A Systems Approach*, Sections 4-6 through 4-8

Objectives:
After performing this experiment, you will be able to:
1. Compute the parameters for an amplifier containing a JFET and a bipolar stage. Construct and test the amplifier.
2. Add a JFET automatic gain control circuit to the amplifier in objective 1.
3. Test a JFET as a constant-current source and as a switch.

Summary of Theory:
JFETs have a nonlinear transfer characteristic as you found in Experiment 11. A change in gate voltage does not produce a linear change in drain current unless the change is small. Some circuits can take advantage of this nonlinear characteristic. For others, the high input resistance offers a significant advantage.

The analysis of JFET amplifiers is complicated by parameter variations between transistors as well as temperature sensitivity. The accurate computation of gain, for example, is dependent on knowing the transconductance, g_m, of the transistor. This value can have a spread of a factor of 5 between transistors of the same type. In addition the gain is small. Bipolar transistors, which are more predictable and have higher gain, are often used in combination with FETs to form an amplifier that takes advantage of both types of transistors.

In this experiment, you will test four applications for JFETs (although there are many others). The first is as a high-resistance input stage to an amplifier that contains a bipolar transistor stage. Since you have completed your study of bipolar transistors, the analysis of this amplifier is straightforward. Next, we exploit the ohmic region of a JFET to provide a variable resistance that can control the gain of the same amplifier. The Q-point is set to $V_{DS} = 0$; no voltage is supplied to the drain! The gate voltage causes the channel resistance to vary.

Two more applications for FETs are then investigated. As a current source, the JFET finds applications in current limiting and waveshaping. Finally, you will add a switching circuit to the current source to change the waveform. JFET switching circuits are normally on when the gate voltage is zero and can be turned off by applying a negative voltage to the gate.

Materials Needed:
Resistors (one of each): 180 Ω, 2.7 kΩ, 3.9 kΩ, 5.1 kΩ, 27 kΩ, 56 kΩ, 1.0 MΩ
Capacitors (one of each): 0.1 μF, 1.0 μF, 10 μF
Transistors: one 2N3904 *npn* transistor, two 2N5458 *n*-channel JFETs (or equivalent)
One LED
One milliammeter 0 - 10 mA

Procedure:

The JFET as a Preamp

1. Compared to bipolar transistors, the major advantage of FETs is the extremely high input resistance. The major disadvantages are the low gain and nonlinearity. To capitalize on the best features of both, FETs are often used in combination with bipolar transistors. Figure 14-1 shows a combination circuit consisting of a high input resistance JFET stage and a relatively high gain CE amplifier. Measure the resistors listed in Table 14-1 and construct the amplifier.

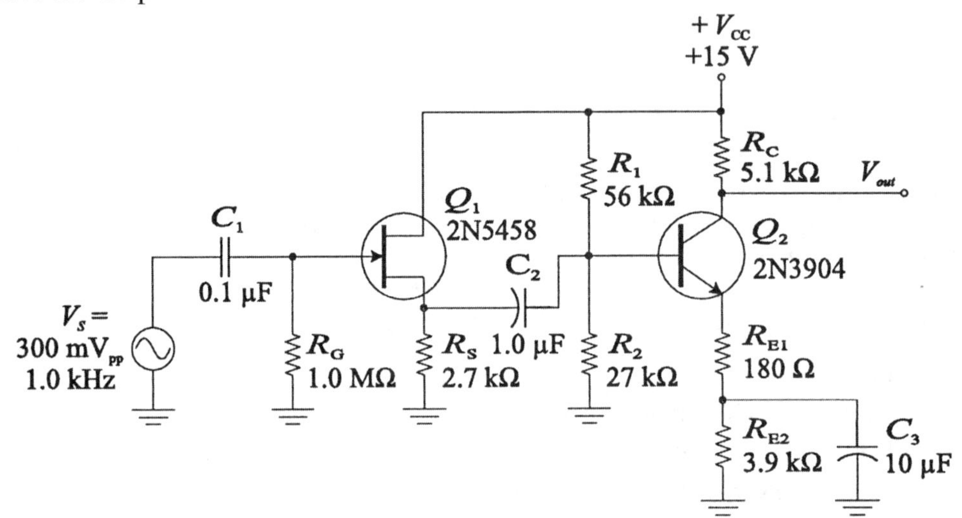

Figure 14-1

Table 14-1

Resistor	Listed Value	Measured Value
R_G	1.0 MΩ	
R_S	2.7 kΩ	
R_1	56 kΩ	
R_2	27 kΩ	
R_{E1}	180 Ω	
R_{E2}	3.9 kΩ	
R_C	5.1 kΩ	

2. Calculate the parameters listed for the amplifier. The parameters for the CE stage have been described in Experiment 8. For calculation purposes, the gain of the common-drain stage is approximately 0.75 (actual value depends on g_m). Use 0.75 to compute the ac base voltage, V_b. Measure and verify that your calculations are reasonable and that the amplifier works as expected. Enter your computed and measured values in Tables 14-2 and 14-3.

Table 14-2

DC Parameter	Computed Value	Measured Value
V_B		
V_E		
I_E		
V_C		
V_{CE}		

Table 14-3

AC Parameter	Computed Value	Measured Value
V_b		
r_e		
$A_{v(Q1)}$	0.75	
$A_{v(Q2)}$		
A_v'		

The JFET as a Variable Resistor

3. The ohmic region of a JFET can be used to provide voltage controlled gain. (A similar idea will be investigated in the For Further Investigation of Experiment 15.) Modify the amplifier you constructed in steps 1 and 2 to include the additional JFET as shown in Figure 14-2. Notice that C_3 is moved from its previous location. Measure the maximum gain with V_{GG} set to 0 V. Then slowly increase V_{GG} while observing the output. Measure the minimum gain when the JFET is turned off (approximately – 5 V). Record your observations in Table 14-4.

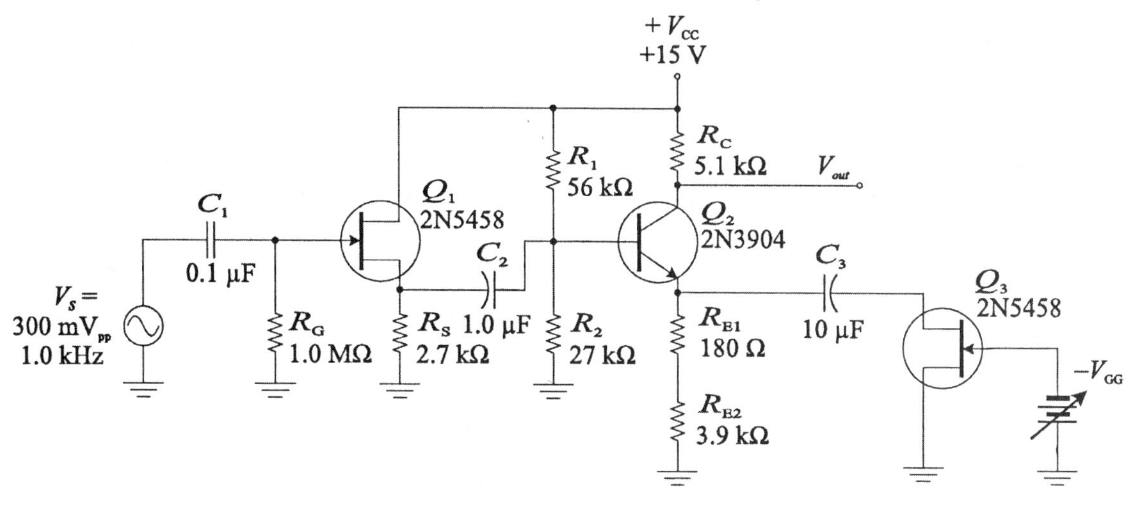

Figure 14-2

Table 14-4

Overall Gain	Measured Value
$A_{v(MIN)}$	
$A_{v(MAX)}$	

The JFET as a Current Source

4. A JFET can be used as a constant-current source of I_{DSS} for applications such as sourcing current to an LED. Build the circuit shown in Figure 14-3. Vary the drain voltage, V_{DD}, between +5 V and +15 V. What is the current? Summarize your observations.

____

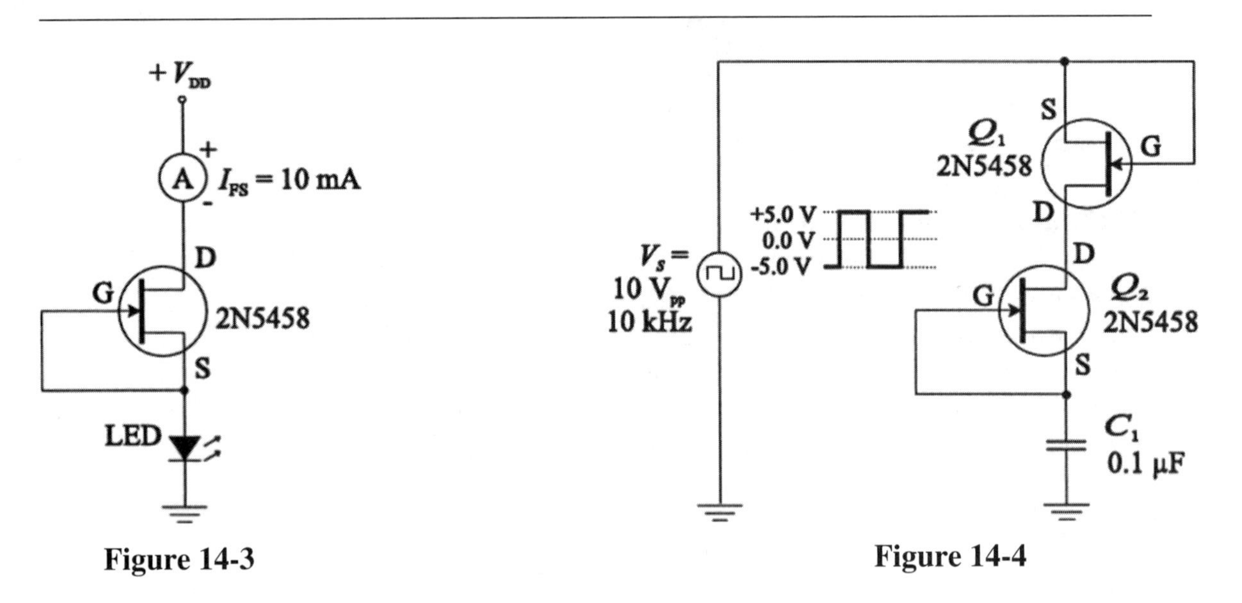

Figure 14-3

Figure 14-4

5. Change the circuit to that in Figure 14-4. Drive the circuit with the signal generator set to a 10 V_{pp} square wave as illustrated. Note that the drains of the JFETs are connected together. While one FET is biased to source I_{DSS}, the other looks like a forward-biased diode - thus the back to back FETs can source current in either direction! Sketch the waveform across the capacitor on Plot 14-1. Label the voltage and time scales on your plot.

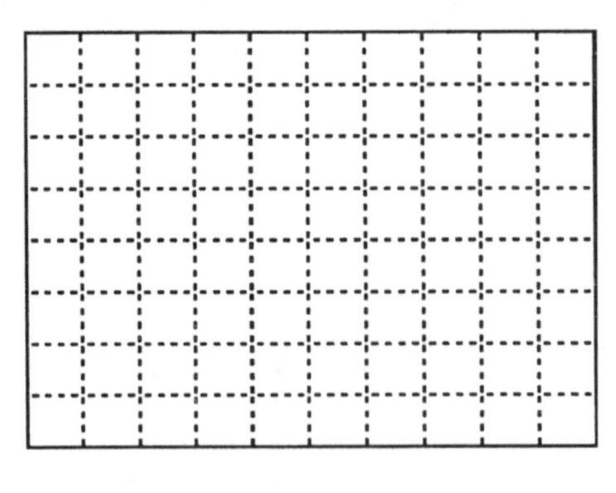

Plot 14-1

6. Vary the amplitude and frequency of the source as you observe the waveform across the capacitor. Can you explain why it becomes nonlinear? Summarize your observations.

____

The JFET as a Switch

7. JFETs are also frequently used in switching applications. The circuit shown in Figure 14-5 uses one FET as a current source, the second as a switch. Consider the circuit, then predict the waveform you will see on the capacitor. Modify the circuit from step 5 (Figure 14-4) and observe the capacitor waveform. Sketch the waveform on Plot 14-2. Label the voltage and time scales.

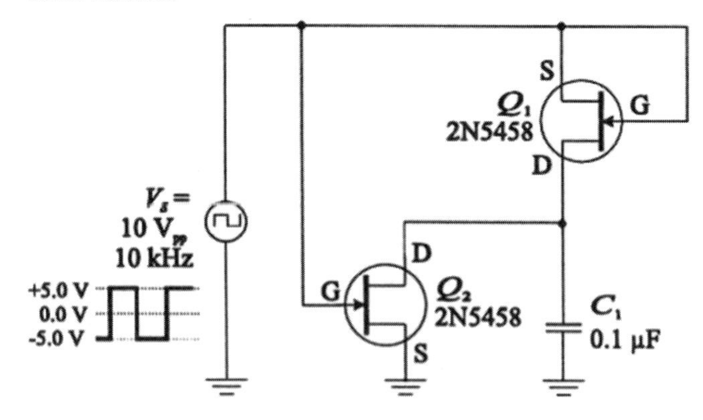

Figure 14-5

Plot 14-2

Conclusion:

Evaluation and Review Questions:

1. What are the advantages of mixing a FET and a bipolar transistor in the same circuit as in Figure 14-1?

2. How is it possible to operate Q_3 (Figure 14-2) with no dc drain voltage? Where is the dc operating point?

3. What change would you make to the circuit in Figure 14-3 to cause half as much current in the LED?

4. (a) From the sketch you drew in step 5, determine the rate that the capacitor was charging and discharging ($\frac{\Delta V}{\Delta t}$).

   (b) The rate of change of voltage times the capacitance is equal to the charging current for a capacitor, as given by the equation

$$i = C\frac{\Delta V}{\Delta t}$$

   Using the rate of change in part (a) and the capacitance in Figure 14-4, show that the charging current for the capacitor is approximately I_{DSS}.

5. Explain how the circuit in Figure 14-5 works.

For Further Investigation:

The amplifier in this experiment was connected in *cascade,* which means that stages are connected in series. An interesting variation is to connect it in *cascode*. In a cascode configuration, the drain of the FET supplies ac signal *current* to the emitter of the next stage. The second stage is a common-base amplifier. While the overall gain is not high, the cascode connection has a much higher frequency response than the cascade connection. Construct the cascode amplifier shown in Figure 14-6. Set the signal generator for a 300 mV_{pp} input signal. Test the dc and ac parameters and submit a short report describing your findings.

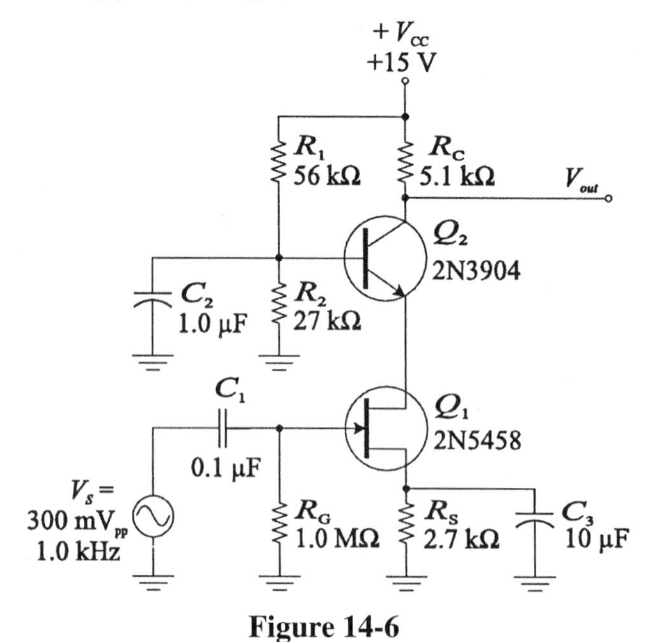

Figure 14-6

Name ______________
Date ______________
Class ______________

15 Multistage Amplifiers

Reading:
Floyd and Buchla, *Analog Fundamentals: A Systems Approach*, Sections 5-1 through 5-4

Objectives:
After performing this experiment, you will be able to:

1. Construct a two-stage transistor amplifier and measure the dc and ac parameters including the input resistance, output resistance, voltage gain, and power gain.
2. *For Further Investigation:* Add automatic gain control to the amplifier in objective 1 and plot the gain characteristic of the amplifier as a function of the amplitude of the input signal.

Summary of Theory:
A single stage of amplification is often not enough for a particular application. The overall gain can be increased by using more than one stage. Frequently, the first stage is a low-noise voltage amplifier which is followed by additional voltage or power amplification. An example of this was given in the For Further Investigation section of Experiment 9 where a CE stage was followed by a dc coupled CC stage to form a two stage amplifier.

Sometimes the input signal is not a fixed quantity. An example is in communication receivers, where the signal strength varies for a number of reasons such as the signal path, weather conditions, time of day or night, etc. In cases like this, a compensation circuit, called automatic gain control (AGC), is useful to decrease the gain when the signal rises and increase the gain when the signal falls. In the For Further Investigation section of this experiment, you will add a field-effect transistor to the circuit to provide AGC.

The two-stage linear amplifier is shown in Figure 15-1. It uses two common emitter (CE) circuits, with the *pnp* and *npn* transistors connected in a cascade amplifier. R_A and R_B are not considered part of the amplifier, but are only used to attenuate the input signal from the function generator by a known factor. To analyze the amplifier, start with the dc parameters. Use measured values of components in your calculations. The steps to solve for the dc parameters for this amplifier are:

1. Mentally remove (open) capacitors from the circuit since they appear open to dc. Solve for the base voltage, V_B of Q_1. By inspection, the dc base voltage is zero; however if the resistors are not equal, the base voltage can be found by applying the voltage-divider rule and the superposition theorem to R_1 and R_2.
2. Add the 0.7 V forward-bias drop across the base-emitter diode of Q_1 from V_B to obtain the emitter voltage, V_E, of Q_1.
3. Find the voltage across the emitter resistors and apply Ohm's law to solve for the emitter current, I_E, of Q_1.

4. Assume the emitter current (step 3) is equal to the collector current, I_C, of Q_1. Find the voltage across R_{C1} and the voltage drop across Q_1. Solve for the dc voltage drop across this equivalent resistance to find the voltage at the collector of Q_1.
5. Compute the base voltage of Q_2 by applying the superposition theorem and voltage divider rule to R_3 and R_4. Subtract 0.7 V from the base voltage of Q_2 to find the emitter voltage of Q_2. Find the voltage across the emitter resistors and apply Ohm's law to determine the emitter current in Q_2.
6. Assume the emitter current (step 5) is equal to the collector current, I_C, of Q_2. Find the voltage across R_{C2} and the voltage drop across Q_2.

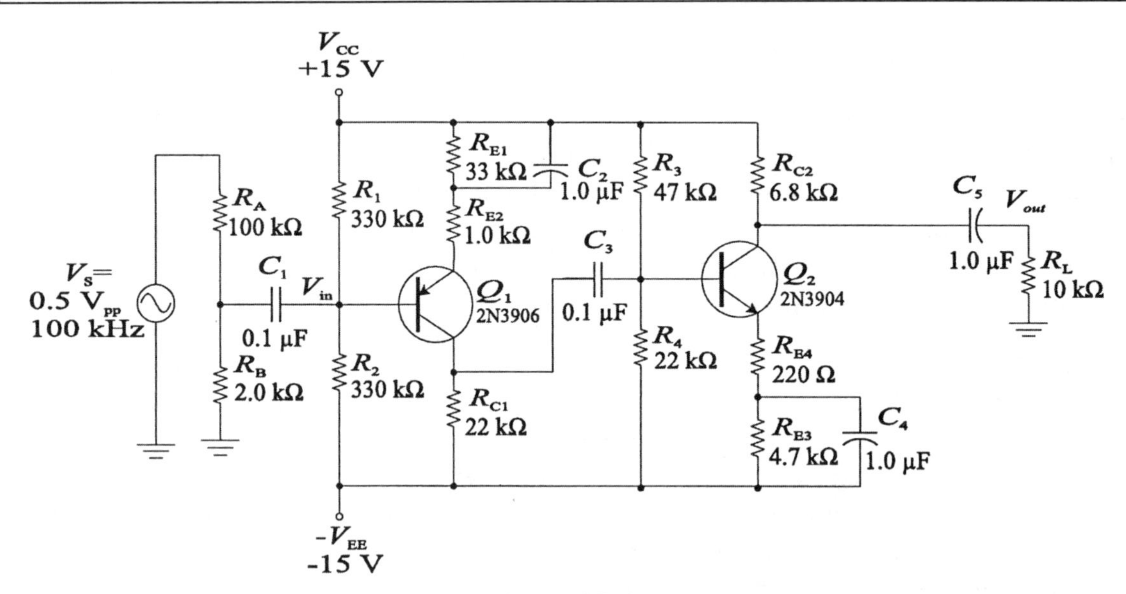

Figure 15-1

The **ac** parameters for the amplifier can now be analyzed. The ac analysis steps are:

1. Replace all capacitors with a short. The ac resistance in the emitter circuit includes the unbypassed emitter resistor and the ac resistance of the transistor. Compute the ac emitter resistance of each transistor, r_e', from the equation:

$$r_e' = \frac{25\text{ mV}}{I_E}$$

2. Compute the input and output resistance of Q_1. The input resistance includes the bias resistors in parallel with the ac resistance of the emitter circuit reflected into the base circuit. The output resistance is simply the value of the collector resistor.

$$R_{in\ (Q1)} = R_1 \parallel R_2 \parallel \left\{\beta_{ac}\left(r_e' + R_{E2}\right)\right\}$$

3. Compute the input and output resistance of Q_2. As before, the input resistance includes the bias resistors and the ac emitter resistance reflected to the base circuit. The output resistance is again the collector resistor.

$$R_{in\ (Q2)} = R_3 \parallel R_4 \parallel \left\{\beta_{ac}\left(r_e' + R_{E4}\right)\right\}$$

4. Compute the unloaded gain, $A_{v(NL)}$, of each stage. The unloaded voltage gain for the common-emitter transistors can be written:

$$A_{v\,(NL)} = \frac{V_{out}}{V_{in}} = \frac{I_c R_C}{I_e \left(r_e' + R_{e\,(ac)}\right)} = \frac{R_C}{\left(r_e' + R_{e(ac)}\right)}$$

5. Compute the overall gain of the amplifier. It is easier to calculate the voltage gain of a multistage amplifier by computing the *unloaded* voltage gain for each stage, then including the loading effect by computing voltage dividers for the output resistance and input resistance of the following stage. This idea is illustrated in Figure 15-2. Each transistor is drawn as an amplifier consisting of an input resistance, R_{in}, an output resistance, R_{out}, along with its unloaded gain, A_v. Then, the overall loaded gain, A_v', of this amplifier can be found by:

$$A_v' = A_{v\,1}\left(\frac{R_{in\,2}}{R_{out\,1} + R_{in\,2}}\right)A_{v\,2}$$

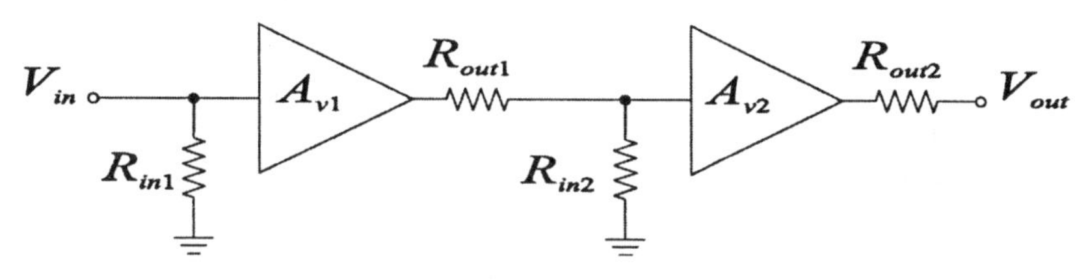

Figure 15-2

Note that if a load resistor was added across the output, an additional voltage divider consisting of the output resistance of the second stage and the added load resistor is used to compute the new gain.

Materials Needed:

One 2N3904 *npn* transistor (or equivalent)
One 2N3906 *pnp* transistor (or equivalent)
Capacitors: two 0.1 μF, three 1.0 μF
Resistors: one of each: 220 Ω, 1.0 kΩ, 2.0 kΩ, 4.7 kΩ, 6.8 kΩ, 10 kΩ, 33 kΩ, 47 kΩ
Resistors: two of each 22 kΩ, 100 kΩ, 330 kΩ
One 100 kΩ potentiometer

For Further Investigation:

One MPF102 *n*-channel JFET
One 1N914 signal diode (or equivalent)
Resistors: one 100 Ω, one 220 kΩ
Capacitors: one additional 1.0 μF capacitor

Procedure:

1. Measure and record the values of the resistors listed in Table 15-1. Compute the dc parameters for the amplifier listed in Table 15-2.

Table 15-1

Resistor	Listed Value	Measured Value
R_A	100 kΩ	
R_B	2.0 kΩ	
R_1	330 kΩ	
R_2	330 kΩ	
R_{E1}	33 kΩ	
R_{E2}	1.0 kΩ	
R_{C1}	22 kΩ	
R_3	47 kΩ	
R_4	22 kΩ	
R_{E3}	4.7 kΩ	
R_{E4}	220 Ω	
R_{C2}	6.8 kΩ	
R_L	10 kΩ	

Table 15-2

DC Parameter	Computed Value	Measured Value
$V_{B(Q1)}$		
$V_{E(Q1)}$		
$I_{E(Q1)}$		
$V_{C(Q1)}$		
$V_{CE(Q1)}$		
$V_{B(Q2)}$		
$V_{E(Q2)}$		
$I_{E(Q2)}$		
$V_{C(Q2)}$		
$V_{CE(Q2)}$		

2. Construct the two stage bipolar transistor amplifier shown in Figure 15-1. The function generator should be turned off. Measure and record the dc voltages listed in Table 15-2. Your measured and computed values should agree within 10%.

3. Compute the ac parameters listed in Table 15-3. The gains for each stage are not loaded. The output resistance of Q_1 ($R_{out(Q1)}$) is simply the collector resistor; the input resistance of Q_2 is determined by the procedure given in the Summary of Theory.

4. Compute the overall gain of the amplifier using the computed gains from Table 15-3 and the input and output resistance between the stages (see step 5 of the ac analysis in the Summary of Theory). Enter the computed overall gain, A_v', on the first line of Table 15-4. Using this value, compute the expected output voltage; enter the computed output voltage on the last line of Table 15-4.

5. Connect the function generator voltage to the divider composed of R_A and R_B as shown in Figure 15-1. (*Note*: The purpose of these resistors is to attenuate the generator signal by a known amount; they will not be considered part of the amplifier.) Turn on the function generator and set V_s for a 0.5 V_{pp} sine wave at 100 kHz. (Check voltage and frequency with the oscilloscope.) The ac base voltage of Q_1 is V_{in}; it is shown as 10 mV_{pp} (based on the voltage divider). Measure the ac signal voltage at the amplifier's output ($V_{out(Q2)}$) and record the value on the last line of Table 15-4. Then use V_{in} and V_{out} to find the measured overall gain, A_v'. Record the measured overall gain in the first line of Table 15–4.

Table 15-3

AC Parameter	Computed Value
$r'_{e\,(Q1)}$	
$r'_{e\,(Q2)}$	
$A_{v\,(NL)(Q1)}$	
$A_{v(NL)(Q2)}$	
$R_{out(Q1)}$	
$R_{in(Q2)}$	

Table 15-4

AC Parameter	Computed Value	Measured Value
A'_v		
$R_{in(Q1)}$		
$R_{out(Q2)}$		
$V_{in(Q1)}$	10 mV	
$V_{out(Q2)}$		

6. The measurement of the total input resistance, $R_{in(tot)}$, is done indirectly by the method shown in Experiment 8 (step 6) using a variable test resistor. The method is repeated here. The output signal (V_{out}) set by V_s to a convenient level with the amplifier operating normally (no clipping or distortion). The output is observed (with an oscilloscope) and the amplitude noted. A variable test resistor (R_{test}) is then inserted in series with the source as shown in Figure 15-3. Because of the higher input resistance of this amplifier, a 100 kΩ test resistor is shown. The resistance of R_{test} is increased until V_{out} drops to one-half the value prior to inserting R_{test}. With this condition, $V_{in} = V_{test}$ and $R_{in(tot)}$ must be equal to R_{test}. R_{test} can then be removed and measured with an ohmmeter. The total input resistance, $R_{in(tot)}$, is the same as the ac input resistance to Q_1 ($R_{in(Q1)}$). Using this method, measure $R_{in(Q1)}$ and record the result in Table 15-4.

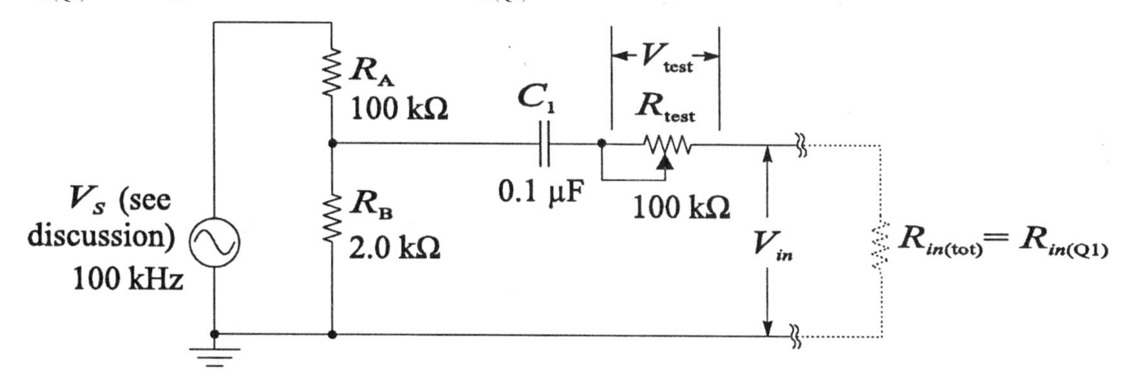

Figure 15-3 Measurement of $R_{in(tot)}$.

7. In this step you will measure the output resistance of the amplifier. The computed output resistance is the same as R_{C2}, the load resistor of Q_2. You can measure the output resistance of any amplifier by measuring the loading effect caused by adding a load resistor. It is not necessary that the load resistor be equal to the output resistance. Consider the model of an amplifier shown in Figure 15-2. Assume that you want to indirectly measure R_{out2}. Think of the amplifier as a Thevenin source with the Thevenin resistance represented by the output resistance. You can find this resistance by measuring the unloaded output voltage and the new output voltage when a known load resistor is placed on the output. Use this idea to develop the equation for the output resistance of the amplifier. Then make the measurements and solve for the output resistance.

Conclusion:

Evaluation and Review Questions:

1. Explain why it is necessary to measure the input resistance indirectly.

2. When you calculated the ac parameters, you were instructed to consider the capacitors as *shorts*. Under what conditions is this assumption *not* warranted?

3. For the circuit in the experiment, the output resistance of both transistors was considered to be the individual collector resistor. Explain.

4. Assume you wanted to use base bias for Q_2 as shown in the circuit of Figure 15-1.

   (a) Explain the changes you would make to the amplifier to accomplish this.

   (b) What disadvantage would result from this change?

5. What is the phase between the input and output signal? Explain your answer.

For Further Investigation:

Add the Automatic Gain Control (AGC) shown in Figure 15-4. The AGC circuit consists of transistor Q_3, diode D_1, capacitors C_2 and C_6 and resistors R_5 and R_6. (Note that C_2, which served as a bypass capacitor in Figure 15-1, is now a coupling capacitor for the AGC circuit.) This AGC will limit the gain moderately because control is applied to only one stage, whereas in many applications, the gain control voltage is applied to several stages. Test the gain with the input signals listed in Table 15-5. The input signal is computed from the function generator setting and assumed to be 2% of the function generator value due to the input voltage divider. Graph the overall gain versus the input signal amplitude with AGC on Plot 15-1. Label your graph.

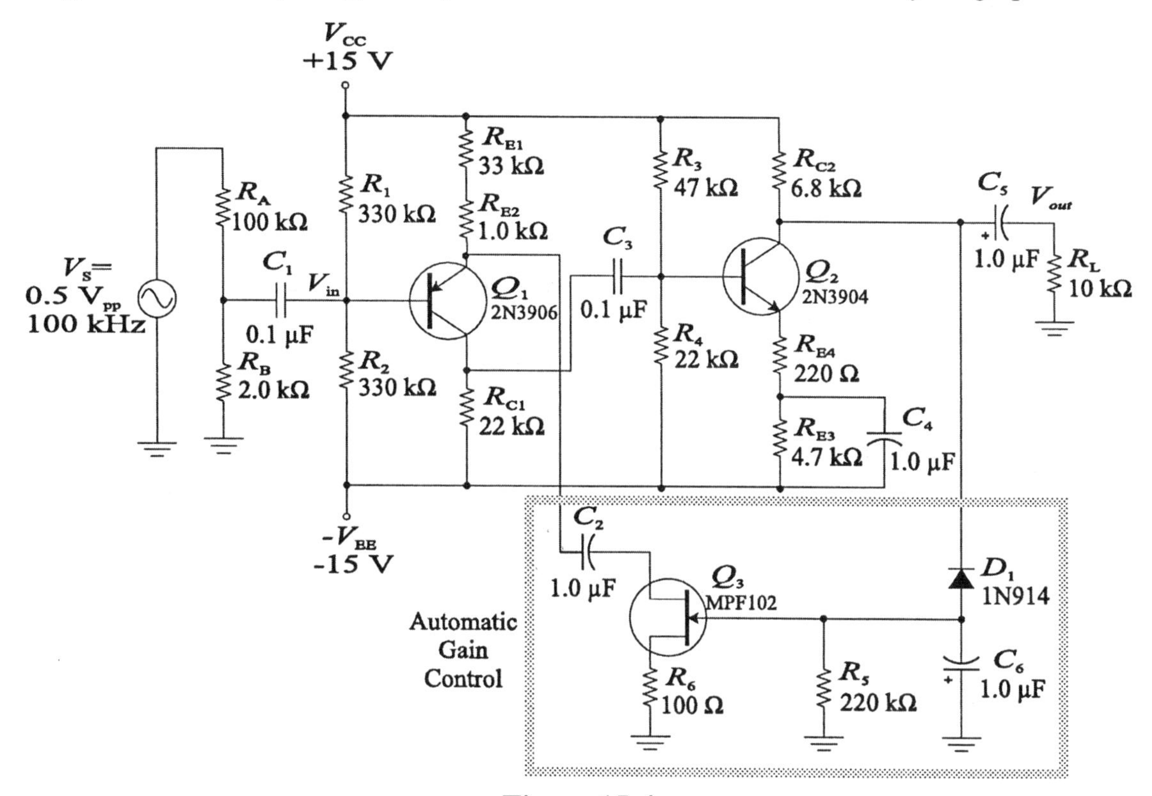

Figure 15-4

Table 15-5

Generator Setting	V_{in}	V_{out}	A_v
0.5 V	10 mV		
1.0 V	20 mV		
2.0 V	40 mV		
4.0 V	80 mV		
6.0 V	120 mV		
8.0 V	160 mV		
10.0 V	200 mV		
20.0 V	400 mV		

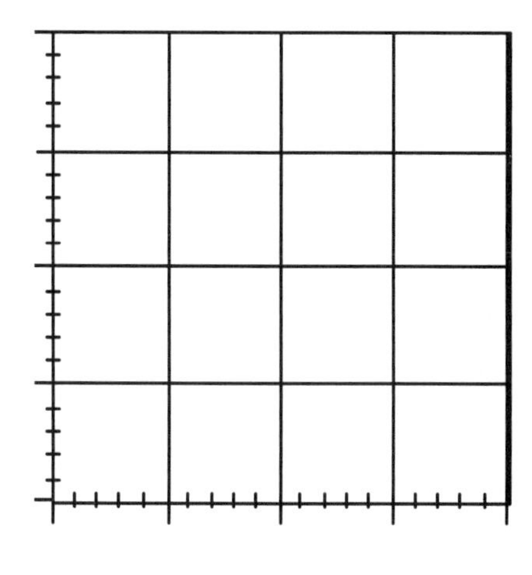

Plot 15-1

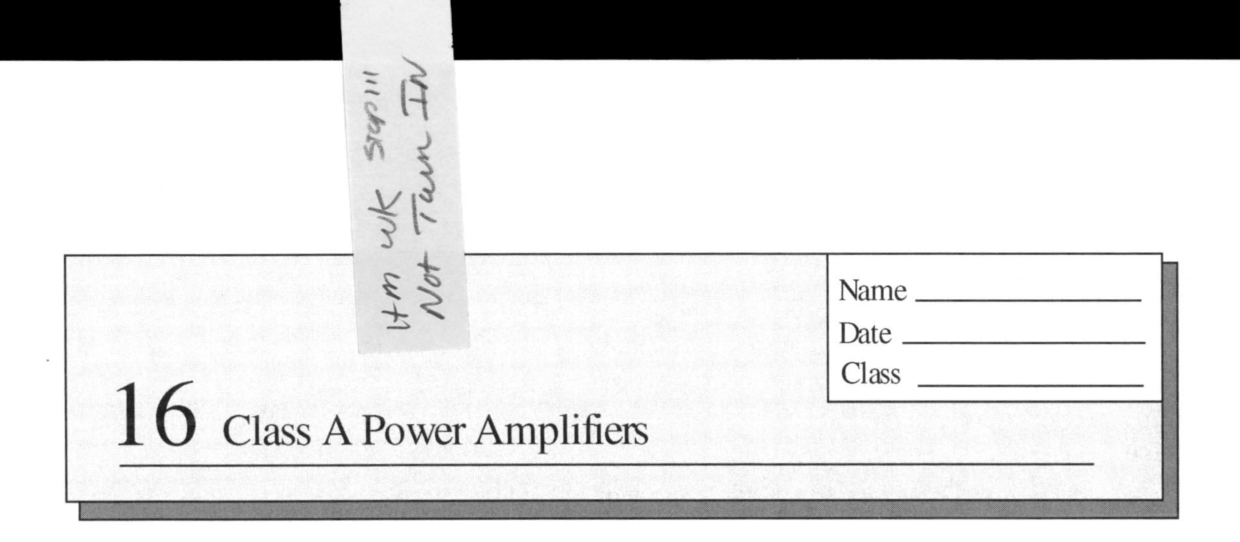

16 Class A Power Amplifiers

Reading:
Floyd and Buchla, *Analog Fundamentals: A Systems Approach*, Section 5-5

Objectives:
After performing this experiment, you will be able to:
1. Calculate the dc and ac parameters for a multistage class A power amplifier.
2. Construct the circuit in Objective 1 and measure the dc and ac parameters.
3. Predict and test the effect of troubles in a two-stage amplifier.
4. Determine the ac power dissipated in the load (speaker) and the power gain of the circuit.

Summary of Theory:
The common-emitter (CE) amplifier provides high voltage gain with moderate input resistance whereas the common collector (CC) amplifier provides current gain and low output resistance. Combining the two amplifiers gives the advantages of each, allowing the amplifier to drive a relatively low resistance load such as a speaker.

In all amplifiers, some power from the supply is wasted – that is, it does not show up as signal power in the load. In class A power amplifiers, the transistor is biased on at all times, causing power to be dissipated in the transistor, even when no signal is present. Because of this, class A amplifiers are not as efficient as class B designs. For low power applications, this reduced efficiency is not a major problem. Further, the power dissipated in the transistor is highest when no signal is present, so it is simple to compute the worst case power dissipated in the transistor – it is simply $V_{CEQ}I_{CQ}$.

In this experiment, you will combine a CE and CC amplifier to form an amplifier that will be used for driving a small speaker. Most speakers are low resistance devices, requiring the driving amplifier to have a low output resistance. Because the CE amplifier typically has relatively high output resistance, a Darlington CC amplifier is selected to minimize the loading effect. In the For Further Investigation section, you can complete the amplifier by adding a common base (CB) driver and microphone for a small intercom system. As in earlier experiments, you should analyze the operation of the amplifier by first computing the dc parameters. After analyzing the dc conditions, the ac parameters for the amplifier are evaluated and power input, power output, and power gain are calculated. The analysis steps are reviewed in the experiment.

Materials Needed:
Resistors: one 22 Ω (2 W), one 100 Ω, one 560 Ω, one 4.7 kΩ, two 10 kΩ, one 22 kΩ, one 56 kΩ
Capacitors: one 0.22 μF, one 1.0 μF, one 10 μF, two 100 μF
Transistors: two 2N3904, one SK3024 (or equivalent) with heat sink
One small 8 Ω speaker

For Further Investigation:

One additional 2N3904 transistor
One low resistance microphone (a small speaker can be substituted)
Resistors as specified by student

Procedure:

1. Measure and record the values of the resistors listed in Table 16-1.

Table 16-1

Resistor	Listed Value	Measured Value
R_1	56 kΩ	
R_2	10 kΩ	
R_{E1}	100 Ω	
R_{E2}	560 Ω	
R_C	4.7 kΩ	
R_3	10 kΩ	
R_4	22 kΩ	
R_{E3}	22 Ω	

2. Using the methods described in Experiment 7 and the text, compute the parameters listed in Table 16-2 for the common-emitter stage (Q_1) of the amplifier shown in Figure 16-1. The voltage divider is relatively stiff, so you can use the simplifying assumption that the base voltage for Q_1 is determined directly from the voltage-divider equation. Compute the unloaded and loaded gain. The unloaded gain is found by assuming the coupling capacitor between the CE and CC stages is open. Enter your computed values in Table 16-2.

Table 16-2

Parameters for CE Amp (Q_1)	Computed Value	Measured Value
V_B		
V_E		
I_E		
V_C		
V_{CE}		
r_e'		
$A_{v(NL)}$		
$A_{v(FL)}$		

Table 16-3

Parameters for CC Amp ($Q_{2,3}$)	Computed Value	Measured Value
V_B		
V_E		
I_E		
V_C		
V_{CE}		
r_e'		
$A_{v(NL)}$		
$A_{v(FL)}$		

3. Using the methods described in Experiment 9 and the text, compute the parameters listed in Table 16-3 for the common-collector stage (Q_2 and Q_3) of the amplifier shown in Figure 16-1.

The voltage divider is again relatively stiff. Compute both the unloaded and loaded gain. The unloaded gain is found by assuming the output coupling capacitor, C_4, is open. Enter your computed values in Table 16-3.

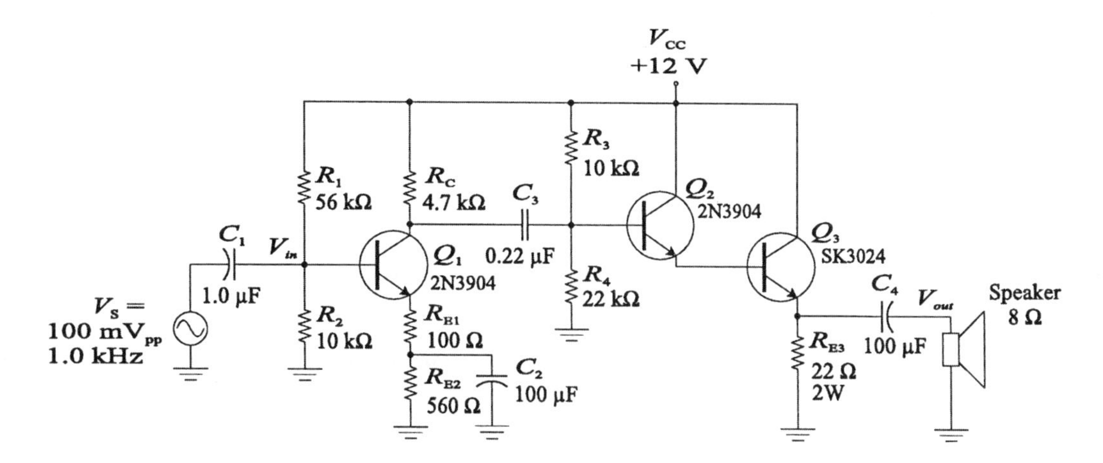

Figure 16-1

4. Construct the circuit shown in Figure 16-1. Q_3 is a power transistor and needs a heat sink. Measure and record the parameters that you computed in steps 2 and 3 in Tables 16-2 and 16-3. When measuring gain, it is important that no distortion or clipping can be observed on the output signal. The unloaded gain of the CE stage is measured by disconnecting the coupling capacitor between the stages, and finding the ratio of the peak-to-peak collector voltage to the peak-to-peak base voltage. To measure the gain of the CC stage, replace the first coupling capacitor and find the ratio of the peak-to-peak output voltage across R_{E3} to the peak-to-peak input base voltage on Q_2. For the unloaded gain, open C_4.
   For the ac parameters, measured and computed values should agree within 10%.

5. In this step, you will determine the power input, load power, and power gain of the amplifier. Enter all values in Table 16-4.

   (a) Enter the speaker's resistance, R_L, in Table 16-4. The value should be marked on the speaker.

   (b) Compute the input resistance of the amplifier, R_{in}.

   (c) While observing the peak-to-peak output voltage on the speaker, increase the input signal to the onset of clipping. Then measure the maximum undistorted output voltage, V_{out}. Convert the reading to rms voltage.

   (d) Without changing the generator, measure the rms input voltage, V_{in}.

   (e) Compute the power to the load, P_L, from the equation, $P_L = \dfrac{V_{out}^2}{R_L}$

   (f) Compute the input power, P_{in}, from the equation, $P_{in} = \dfrac{V_{in}^2}{R_{in}}$

   (g) Compute the power gain, A_p, for the circuit from the ratio of the load to input power.

Table 16-4

Quantity	Value
Load resistance, R_L	
Input resistance, R_{in}	
Output rms voltage, V_{out}	
Input rms voltage, V_{in}	
Load power, P_L	
Input power, P_{in}	
Power gain, A_p	

6. Predict the effects of each trouble listed in Table 16-5 on the amplifier's performance. After predicting the effect of the trouble on the amplifier, then insert the trouble into the circuit and test your prediction. Summarize the results in the space provided in Table 16-5.

Table 16-5

Trouble	Effect on Circuit
R_{E1} switched with R_{E2}	
R_C open	
$C_3 = 10\ \mu F$	
$V_{CC} = 6$ V	
R_4 open	
$C_4 = 10\ \mu F$	
Q_2 has open collector	

Conclusion:

Evaluation and Review Questions:

1. What is the advantage of the Darlington arrangement for the CC amplifier in this experiment?

2. Explain why the loaded and unloaded gain is different for the CE amplifier.

3. A disadvantage of a class-A power amplifier is that it dissipates power from the supply even when no signal is present. From the measurements taken in this experiment, estimate:
   (a) the power supplied by the dc source with no ac input.

   (b) power dissipated in Q_3 when no ac input.

4. From the measurements taken in this experiment, compute the Q-point for Q_3.

   (a) $V_{CEQ} =$

   (b) $I_{CQ} =$

5. Assume you needed to modify the circuit so that you could use a *pnp* transistor for Q_1. The power supply cannot be changed. What changes would you make to the circuit?

For Further Investigation:

A common-base amplifier (such as the one shown in Figure 3-43 of the text) can be driven with a *low* resistance microphone forming the input stage for the circuit used in this experiment to complete an intercom system. (High resistance microphones will be loaded too much by a CB preamp.) You will also need gain adjustment. This can be accomplished by adding a potentiometer that has no effect on the dc conditions but changes the ac resistance of the CE amplifier (how?). Draw a new schematic showing the addition of the CB amplifier and the gain control. Then construct and test the circuit. Note: a small speaker can be used as a low resistance microphone.

Name ______________
Date ______________
Class ______________

17 Class B Push-Pull Amplifiers

Reading:
Floyd and Buchla, *Analog Fundamentals: A Systems Approach*, Section 5-6

Objectives:
After performing this experiment, you will be able to:
1. Construct a push-pull amplifier driven by a common-emitter voltage amplifier.
2. Measure performance characteristics of the circuit constructed in objective 1.

Summary of Theory:
The efficiency of a power amplifier is the ratio of the average signal power delivered to the load to the power drawn from the supply. Amplifiers that conduct continuously are called *class A* amplifiers and are not particularly efficient (typically less than 25%) as you observed in Experiment 16. For small signals, this isn't important, but when a larger amount of power must be delivered, *class B* amplifiers are much more efficient. In class B operation, a transistor is on during 50% of the cycle (half-wave operation). By combining two transistors that alternately conduct on positive and negative half-cycles of the input waveform, a very efficient amplifier can be made. This type of amplifier is called a *push-pull* amplifier. An example, using common collector amplifiers, is shown in Figure 17-1(a). The key to its high efficiency is the fact that the circuit dissipates very little quiescent (standby) power because both transistors are off when no signal is present.

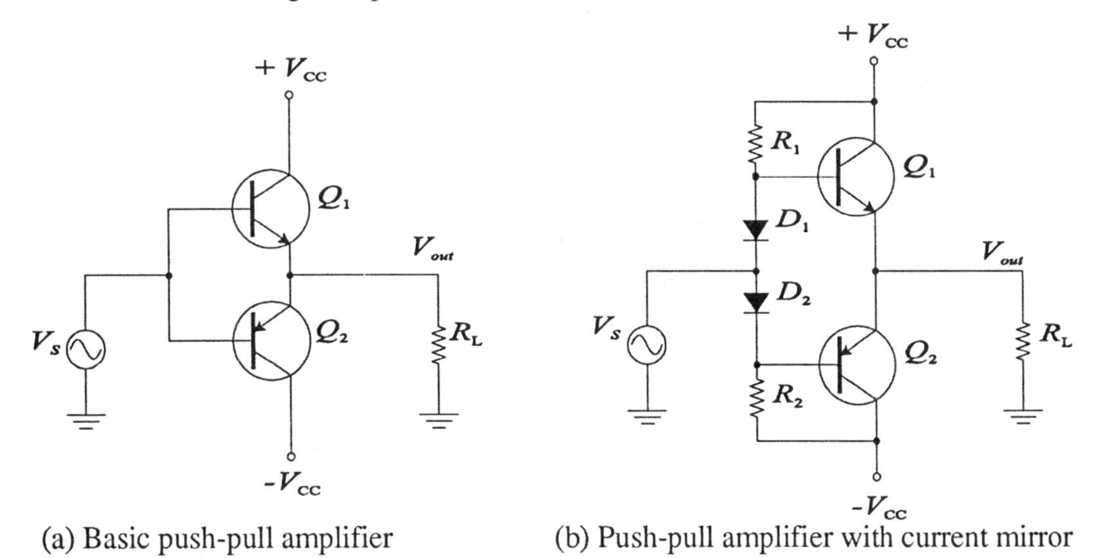

(a) Basic push-pull amplifier (b) Push-pull amplifier with current mirror

Figure 17-1

The push-pull circuit shown uses complementary (*npn* and *pnp*) transistors. One transistor is active for the positive part of the signal (*pushing*), while the other is active for the negative portion of the signal (*pulling*).

A problem with this basic circuit is that the output signal is one diode drop behind the input. This is because the signal must overcome the base-emitter diode drop for each transistor before it will conduct. The output signal follows the input except for the 0.7 V diode drop on both the positive and negative excursion. This causes distortion on the output called *crossover* distortion. Crossover distortion can be eliminated by using diodes to bias the transistors into slight conduction as illustrated in Figure 17-1(b). This type of bias is called *diode current mirror bias*. The forward-biased diodes will each have approximately the same 0.7 V drop as the base-emitter junction. If the diode is matched to the transistor's base-emitter diode, the current in the collector circuit is equal ("mirrors") the current in the diode.

In addition to eliminating cross-over distortion, the current mirror offers another advantage. If the temperature increases, the output current will tend to increase. If the diodes are identical to the base-emitter junction, and in the same thermal environment, any thermal change will tend to be compensated by the diodes, thus maintaining stable bias.

Materials Needed:

Resistors: one 330 Ω, one 2.7 kΩ, two 10 kΩ, one 68 kΩ
One 1.0 μF capacitor
Transistors: one 2N3906 *pnp*, two 2N3904 *npn* (or equivalent)
Two 1N914 diodes (or equivalent)
One 5 kΩ potentiometer

For Further Investigation:

One 15 kΩ resistor, one additional resistor to be determined by student

Procedure:

1. Measure and record the resistance of the resistors listed in Table 17-1.

2. Construct the push-pull amplifier shown in Figure 17-2. The amplifier uses the input signal from the generator to bias the transistors on. Set the generator for a 10 V_{pp} sine wave at 1.0 kHz. Be sure there is <u>no</u> dc offset from the generator. The dual positive and negative power supplies offer the advantage of not requiring large coupling capacitors.

Table 17-1

Resistor	Listed Value	Measured Value
R_L	330 Ω	
R_1	10 kΩ	
R_2	10 kΩ	
R_3	68 kΩ	
R_4	2.7 kΩ	

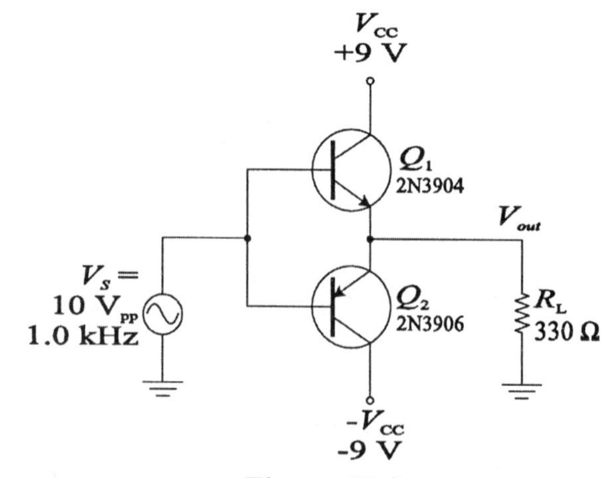

Figure 17-2

3. Sketch the input and output waveforms you observe on Plot 17-1. Show the amplitude difference between the peak input waveform and the output waveform and note the crossover distortion on your plot. Label the plot for voltage and time.

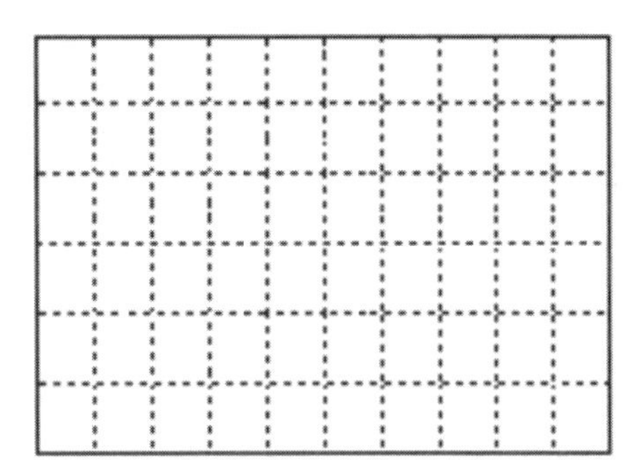

Plot 17-1

4. Add the diode current mirror bias shown in Figure 17-3. Compute the dc parameters listed in Table 17-2. The dc emitter voltage will be 0 V if each half of the circuit is identical and there is no dc offset from the generator. The current in R_1 can be found by applying Ohm's law to R_1. This current is nearly identical to I_{CQ} because of current mirror action.

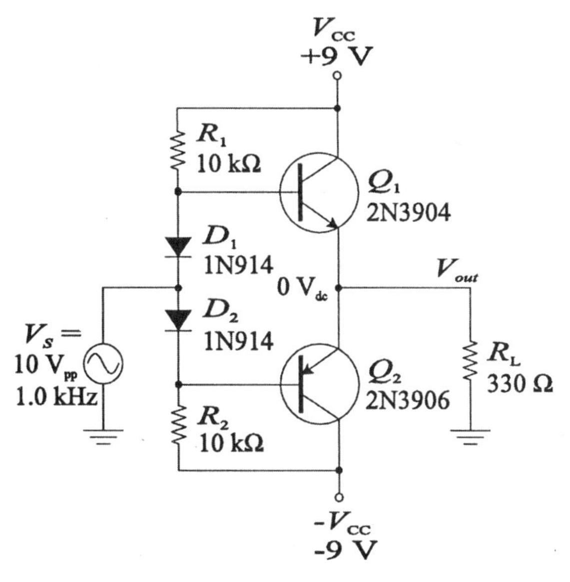

Figure 17-3

Table 17-2

DC Parameter	Computed Value	Measured Value
V_E		
V_{B1}		
V_{B2}		
$I_{R1} = I_{CQ}$		

5. Compute the ac parameters listed in Table 17-3. Compute the maximum undistorted output voltage (V_p) and current (I_p). With dual power supplies, the output can swing nearly to positive and negative V_{CC}. Then compute the peak output current based on the load resistance. The ac power is found by $P_{out} = 0.5I_{p(out)}V_{p(out)}$. By substituting for I_p, the ac power out can also be expressed as:

$$P_{out} = \frac{V_{p(out)}^2}{2R_L}$$

Table 17-3

AC Parameter	Computed Value	Measured Value
$V_{p(out)}$		
$I_{p(out)}$		
$P_{(out)}$		

6. With the signal generator off, apply power, and measure and record the dc parameters listed in Table 17-2.

7. Turn on the signal generator and ensure there is no dc offset. While viewing V_{out} adjust the generator for the maximum unclipped output. Enter $V_{p(out)}$ in Table 17-3.

8. One method for applying a signal to a push-pull amplifier is shown in Figure 17-4. The signal is first amplified by Q_3, a common-emitter amplifier. The quiescent current in the collector circuit produces the same dc conditions as in the circuit of Figure 17-3. The bias adjust allows the dc output voltage to be set to zero to compensate for tolerance variations in the components. Compute the dc parameters listed in Table 17-4. Assume the bias potentiometer is set to 3 kΩ and apply the voltage-divider rule to find $V_{B(Q3)}$. Note that the voltage across the divider string is the difference between $+V_{CC}$ and $-V_{CC}$.

Table 17-4

DC Parameter	Computed Value	Measured Value
$V_{B(Q3)}$		
V_{E3}		
$I_{C(Q3)}$		

Table 17-5

AC Parameter	Computed Value	Measured Value
A_v'		

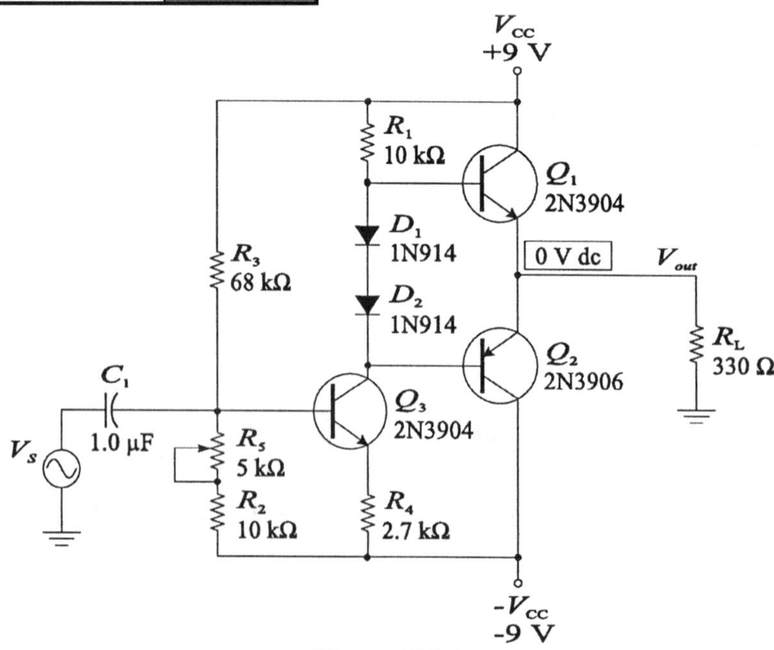

Figure 17-4

9. Compute the voltage gain of Q_3 by taking into account the load presented to the collector circuit of Q_3 by the push-pull amplifier and by dividing by the resistance of the Q_3 emitter circuit. The voltage gain of the push-pull amplifier is nearly 1.0, so the total voltage gain of the amplifier, A_v', is approximately equal to the gain of Q_3. That is,

$$A_v' \cong A_{V\,(Q3)} \cong \frac{R_1 \parallel \left\{\beta_{Q1}\left(R_L + r_{e\,(Q1)}'\right)\right\}}{r_{e\,(Q3)}' + R_4}$$

10. Connect the circuit shown in Figure 17-4. Measure the dc voltage across the load resistor and adjust the bias potentiometer (R_5) for 0 V. Measure and record the voltages listed in Table 17-4. Set the signal generator for the maximum unclipped voltage across the load resistor and measure the total voltage gain. Enter the computed and measured gains in Table 17-5.

Conclusion:

Evaluation and Review Questions:

1. With no signal applied, what power is provided by the power supplies in Figure 17-3? (Remember the current in the diodes is equal to the current in the transistors.)

2. Assume the circuit in Figure 17-3 has a positive half-wave rectified output. What failure(s) could account for this?

3. (a) If one of the diodes in Figure 17-3 shorts, what symptoms will it produce?

   (b) If one of the diodes in Figure 17-3 opens, what symptoms will it produce?

4. In step 10, you found that the total voltage gain was fairly low for the circuit of Figure 17-4. What change to the circuit would you suggest to increase the voltage gain?

5. The bias adjust resistor in Figure 17-4 was chosen to allow a range of bias voltage to Q_3 in order to compensate for variations in components. Compute the minimum and maximum bias voltage based on setting R_5 to its smallest and largest value.

$V_{(bias)min}$ = ___________ $V_{(bias)max}$ = ___________

For Further Investigation:

As you observed in the experiment, a diode current mirror can cause a specific current in a transistor. The technique is used in integrated circuits and in push-pull amplifiers. Another current mirror is illustrated in Figure 17-5(a). Notice that Q_2 is connected as a diode. The current in Q_2 is determined by the drop across R_B which is one diode drop less than the supply voltage. The current in Q_1 is virtually identical to the current in Q_2 because of the identical base-emitter characteristics.

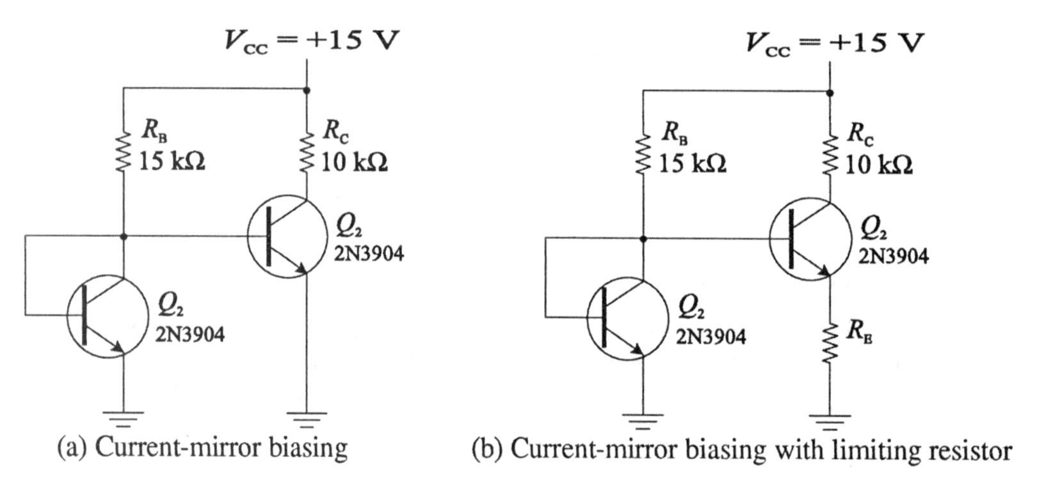

(a) Current-mirror biasing (b) Current-mirror biasing with limiting resistor

Figure 17-5

Current mirrors can be designed to mirror a *smaller* current into a circuit by adding a resistor in the emitter circuit as shown in Figure 17-5(b). The resistor needs to drop a portion of the original base-emitter voltage. A drop of 60 mV in the emitter resistor will cause the collector current of Q_1 to drop by a factor of ten.

Construct the current-mirror circuit shown in Figure 17-5(a). Measure the voltage drop across R_B and determine the collector current in Q_2. Then, using the 60 mV rule, determine the size of the emitter resistor needed to cut the collector current of Q_2 to 10% of its original value. Modify the original circuit by adding the resistor, R_E, that you calculated. Measure the new collector current and compare it with your expected value. Summarize your findings.

Name ______________
Date ______________
Class ______________

18 The Differential Amplifier

Reading:
Floyd and Buchla, *Analog Fundamentals: A Systems Approach*, Section 6-2

Objectives:
After performing this experiment, you will be able to:

1. Compute and measure the basic dc and ac parameters for a differential amplifier.
2. Measure the differential and common-mode gain for a differential amplifier and compute the CMRR′.

Summary of Theory:
The differential amplifier has two inputs and amplifies the *difference* signal applied between the two inputs but ignores any *common* signal applied to the inputs. A difference signal is said to be a normal mode signal, while a signal that is the same on both inputs is said to be a common-mode signal. The ability to amplify a normal mode signal while rejecting a common-mode signal is an important advantage in instrumentation systems, where a small signal may be contaminated by common-mode noise pickup in cabling.

Figure 18-1(a) shows an emitter-biased differential amplifier constructed with two *npn* transistors. The emitter currents for each transistor combine to form the current in the tail resistor, R_T. The signal is applied between the bases and is removed between the collectors. Since there are two inputs and two outputs, this mode of operation is referred to as a double-ended

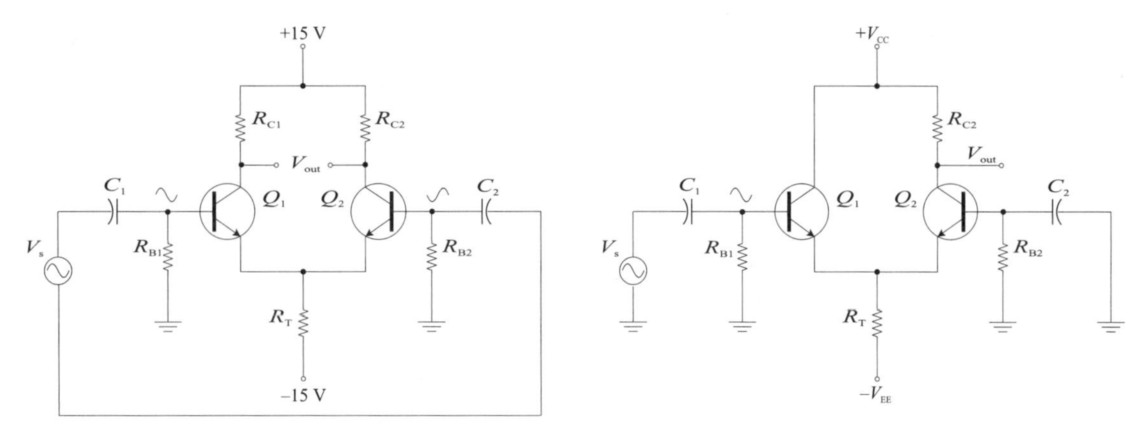

(a) Double-ended input and output

(b) Single-ended input and output

Figure 18-1

input and double-ended output. If the input is applied to only one side, as in Figure 18-1(b), and the output is taken from only one side (output is taken from only one of the collectors), then the mode of operation is referred to as a single-ended input and single-ended output.

The dc conditions for the differential amplifier are computed by finding the tail current and splitting it for the two transistors. Because it uses emitter bias, you can approximate the emitter voltage as –1 V and determine the tail current by applying Ohm's law to R_T. The dc emitter current in each transistor is one-half the tail current.

In this experiment, you will construct and test a differential amplifier. First you will compute the dc conditions; then you will compute the ac conditions. You will then measure the differential-mode and common-mode gains and use these gains to compute the common mode rejection ratio (CMRR). In the For Further Investigation section, you will test a current source biasing circuit and observe how it can improve the CMRR.

Materials Needed:

Resistors: two 100 Ω, two 10 kΩ, one 33 kΩ, two 100 kΩ
Capacitors: two 10 μF
Transistors: two 2N3904 (or equivalent)

For Further Investigation:

one additional 2N3904 transistor
two additional 10 kΩ resistors and one 4.7 kΩ resistor

Procedure:

1. Measure and record the values of the resistors listed in Table 18-1. Best results (for maximum CMRR) can be obtained if R_{B1} and R_{B2} are matched and R_{E1} and R_{E2} are also matched.

Table 18-1

Resistor	Listed Value	Measured Value
R_{B1}	100 kΩ	
R_{B2}	100 kΩ	
R_{E1}	100 Ω	
R_{E2}	100 Ω	
R_T	10 kΩ	
R_{C2}	10 kΩ	

Table 18-2

DC Parameter	Computed Value	Measured Value
V_A	–1 V	
I_T		
$I_{E1} = I_{E2}$		
$V_{C(Q1)}$		
$V_{C(Q2)}$		

2. Construct the differential amplifier shown in Figure 18-2. Because the output is single ended, there is no need for a collector resistor in Q_1. Compute the dc parameters listed in Table 18-2. The voltage at point A is approximated as –1 V for the calculations as shown in the table. I_T is found by applying Ohm's law to the tail resistor, R_T.

3. Measure and record the dc parameters listed in Table 18-2.

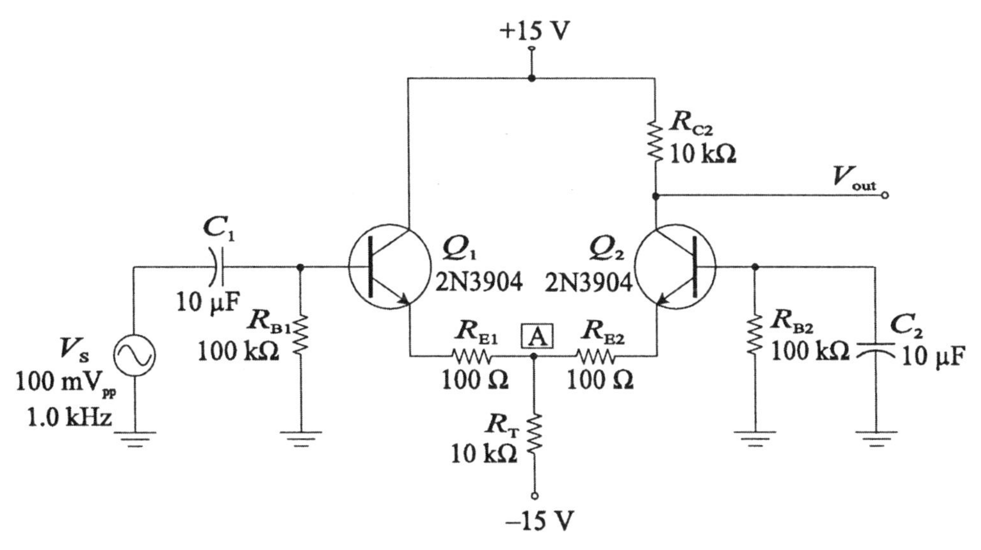

Figure 18-2

4. Compute the ac parameters given in Table 18-3. The differential amplifier can be thought of as a common-collector amplifier (Q_1) driving a common-base amplifier (Q_2). For the single-ended input signal, the overall differential voltage gain is given by:

$$A_{v\,(d)} = \frac{R_{C2}}{2\left(R_{E2} + r_e'\right)}$$

Except for the 2 in the denominator, this equation is equivalent to the gain of a CB amplifier. The reduction by a factor of 2 is due to the attenuation of the signal to point A by the CC amplifier. To compute $R_{in(tot)}$, assume $r_{e\,(Q2)}'$ is in series with R_{E2}, R_{E1}, and $r_{e\,(Q1)}'$. Move this resistance into the base circuit of Q_1 by multiplying by $ß_{Q1}$. This result is then seen to be in parallel with the Q_1 base resistor, R_{B1}. Writing this in equation form:

$$R_{in\,(tot)} = R_{B1} \parallel \left\{ ß_{Q1} \left(r_{e\,(Q1)}' + R_{E1(Q1)} + R_{E1(Q2)} + r_{e\,(Q2)}' \right) \right\}$$

If you don't know $ß_{Q1}$, assume a nominal value of 200 for the 2N3904.

Table 18-3

AC Parameter	Computed Value	Measured Value
$V_{b(Q1)}$	100 mV$_{pp}$	
V_A		
$r_{e\,(Q1)}' = r_{e\,(Q2)}'$		
$A_{v(d)}$		
$V_{c(Q2)}$		
$R_{in(tot)}$		
$A_{v(cm)}$		
CMRR′		

5. Add the single-ended differential mode ac signal as shown in Figure 18-2 (previous page) and measure the ac parameters listed in Table 18-3 (except $A_{v(cm)}$ and CMRR′). To measure $R_{in(tot)}$, note the output voltage, V_c, then add a 33 kΩ test resistor, R_{test}, in series with the input signal and observe the reduced output V_c'. (V_c' should drop to approximately one-half V_c). The measured value of $R_{in(tot)}$ is determined by solving for $R_{in(tot)}$ from the ratio:

$$\frac{V_c'}{R_{in\ (tot)}} = \frac{V_c}{R_{test} + R_{in\ (tot)}}$$

Solving for $R_{in(tot)}$:

$$R_{in\ (tot)} = \left(\frac{V_c'}{V_c - V_c'} \right) R_{test}$$

Notice that if the output drops in half after inserting the test resistor, $R_{test} = R_{in(tot)}$.

6. In this step, you will find the common-mode gain, $A_{v(cm)}$. The common-mode gain can be approximated from the formula:

$$A_{v\ (cm)} \cong \frac{R_C}{2\ R_T}$$

This formula assumes the two sides of the differential amplifier are balanced. Enter the computed common-mode gain in Table 18-3.

7. The common-mode gain is the gain observed when the same signal is applied to both sides of the differential amplifier. Remove the test resistor from step 5 and connect the circuit shown in Figure 18-3. In order to measure the common-mode gain, raise the input signal level from the signal generator until a 1 V_{pp} output is observed. Then measure the ratio of the output to input signal and record the measured $A_{v(cm)}$ in Table 18-3.

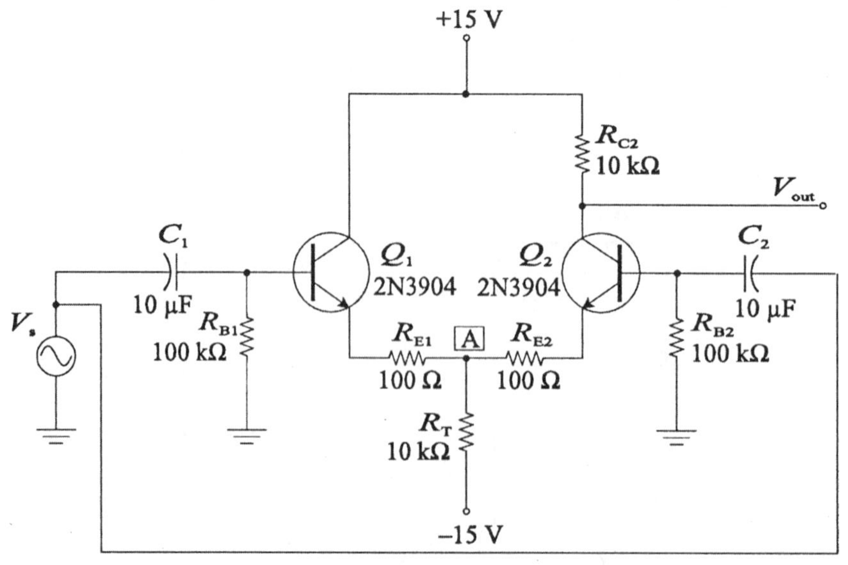

Figure 18-3 Applying a common-mode signal to the differential amplifier.

8. The CMRR′ is 20 times the logarithmic ratio of the absolute value of the ratio of $A_{v(d)}$ to $A_{v(cm)}$, expressed in dB. Expressed as an equation,

$$\text{CMMR}' = 20\log\left|\frac{A_{v\,(d)}}{A_{v\,(cm)}}\right|$$

Compute the CMRR′ for your differential amplifier and record the value in Table 18-3.

Conclusion:

Evaluation and Review Questions:

1. There is no collector resistor for Q_1 in the circuit of Figure 18-2. Why doesn't this have any effect on the dc collector current in either transistor?

2. In step 5, you were directed to measure the input resistance while observing the output voltage, V_c, instead of the input voltage, V_b. Explain what advantage this has to assure a good measurement.

3. Assume the following troubles are associated with the differential amplifier in Figure 18-2. What effect would you expect for each problem on the output signal? Assume only one problem occurs at a time.

   (a) Capacitor C_2 is open:

   (b) Resistor R_{E1} is shorted:

   (c) The transistors have ß's at the opposite extremes of the specified range (one 100, the other 300).

   (d) The negative power supply drops to –10 V.

   (e) The base of Q_2 is shorted to ground.

4. What is the phase relationship of the output signal compared to the input signal?

5. Assume you wanted a higher CMRR′ for the differential amplifier. What improvements to the circuit would you suggest?

For Further Investigation:

The CMRR′ can be improved by reducing the common-mode gain. The internal resistance of a current source is very high. By replacing the tail resistor with a current source, the common-mode gain is reduced and the CMRR′ is made higher. Connect the current-source shown in Figure 18-4 to the circuit in the experiment. Compare the differential and common-mode gains and compute the increased CMRR′. Summarize your results.

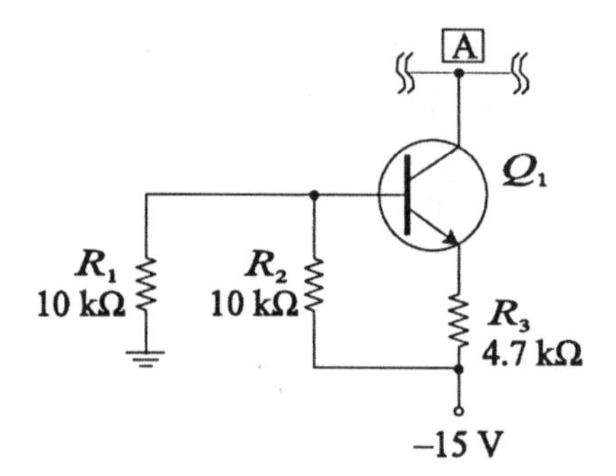

Figure 18-4

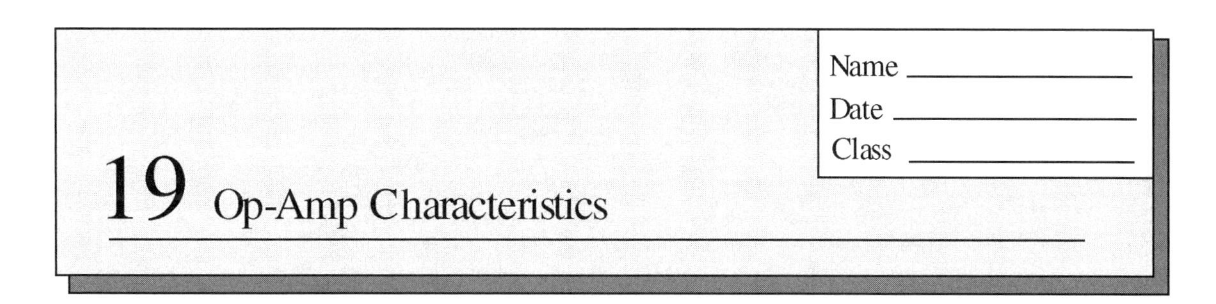

Reading:
Floyd and Buchla, *Analog Fundamentals: A Systems Approach*, Section 6-3

Objectives:
After performing this experiment, you will be able to:

1. Explain the meaning of common op-amp specifications.
2. Use IC op-amp specification sheets to determine op-amp characteristics.
3. Measure the input offset voltage, input bias current, input offset current, CMRR′, and slew rate for a 741C op-amp.

Summary of Theory:
An operational amplifier (*op-amp*) is a linear integrated circuit that incorporates a dc-coupled, high-gain differential amplifier and other circuitry that gives it specific characteristics. The ideal op-amp has certain unattainable specifications, but hundreds of types of operational amplifiers are available, which vary in specific ways from the ideal op-amp. Important specifications are reviewed in the experiment and include open-loop gain, input impedance, output impedance, input offset voltage and current, bias current, and slew rate. Slew rate is defined as the maximum rate of change of the output voltage under large signal conditions and is related to the frequency response. Other important characteristics include CMRR, current and voltage noise level, maximum output current, roll-off characteristics, and voltage and power requirements. The data sheet for a specific op-amp contains these specifications, a description of the op-amp, the device pin-out, internal schematic, maximum ratings, suggested applications, and performance curves.

The input stage of all op-amps is a differential amplifier, so the two inputs are marked with the symbols (+) and (–). These symbols refer to the phase of the output signal compared to the input signal and should be read as noninverting (+) and inverting (–) rather than "plus" or "minus". If the noninverting input is more positive than the inverting input, the output will be positive. If the inverting input is more positive, then the output will be negative.
The symbol for an op-amp is shown in Figure 19-1. The power supply connections are not always shown but must be connected for proper operation. In addition, bypass capacitors are frequently placed near the power connections as shown. Be careful to observe the polarity of electrolytic capacitors when bypassing the power supplies.

+V
Inverting input
Output
Noninverting input
–V

Figure 19-1

Materials Needed:
Resistors: two 100 Ω, two 10 kΩ, two 100 kΩ, one 1.0 MΩ
Two 1.0 μF capacitors
One 741C op-amp
For Further Investigation:
One 10 Ω resistor

Procedure:

1. From the specification sheet for the 741C op-amp (http://www.ti.com), determine the typical and maximum values for each quantity listed in Table 19-1. Record the specified values for $T_A = 25^\circ$C. Note the units that are listed on the right side of the specification sheet.

Table 19-1

Step	Parameter	Specified Value			Measured Value
		Minimum	Typical	Maximum	
2d	Input Offset Voltage, V_{OS}				
3d	Input Bias Current, I_{BIAS}				
3e	Input Offset Current, I_{OS}				
4b	Differential Gain, $A_{v(d)}$				
4c	Common-Mode Gain, A_{cm}				
4d	CMRR′				
5	Slew Rate		0.5 V/μs		

2. In this step, you will measure the input offset voltage, V_{OS}, of a 741C op-amp. The input offset voltage is the amount of voltage that must be applied between the *input* terminals through two equal resistors to give zero *output* voltage. It is a dc parameter.
(a) Measure and record the resistors listed in Table 19-2. R_C is for bias compensation.
(b) Connect the circuit shown in Figure 19-2. Install 1 μF bypass capacitors on the power supply leads as shown. Note the polarities of the capacitors.
(c) Measure the output voltage, V_{OUT}. The input offset voltage is found by dividing the output voltage by the closed loop gain. The closed-loop gain is: $A_{v(cl)} = R_f/R_i + 1$ (assuming a noninverting amplifier for the purpose of the offset calculation).
(d) Record the measured V_{OS} in Table 19-1.

Table 19-2

Resistor	Listed Value	Measured Value
R_f	1.0 MΩ	
R_i	10 kΩ	
R_C	10 kΩ	

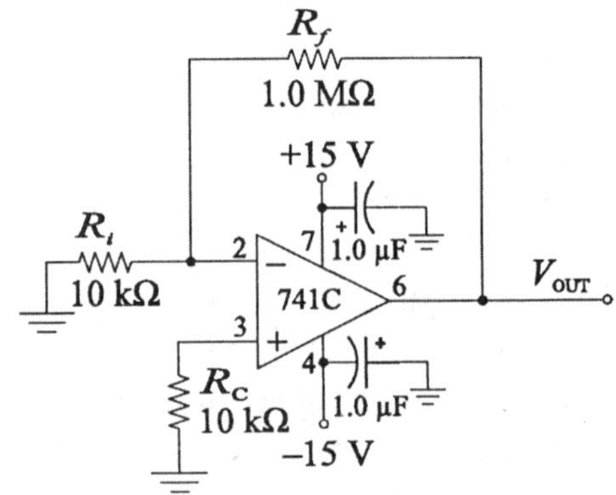

Figure 19-2

3. In this step, you will measure the input bias current, I_{BIAS}, and the input offset current, I_{OS}, of a 741C op-amp. The input bias current is the average of the input currents at each input terminal. The input offset current is a measure of how well these two currents match. The input offset current is the difference in the two bias currents when the output voltage is 0 V. The input bias current and input offset current are dc parameters.
   (a) Measure and record the resistors listed in Table 19-3.
   (b) Connect the circuit shown in Figure 19-3.
   (c) Measure the voltage across R_1 and R_2 of Figure 19-3. Use Ohm's law to calculate the current in each resistor.
   (d) Record the *average* of these two currents in Table 19-1 as the input bias current, I_{BIAS}.
   (e) Record the *difference* in these two currents in Table 19-1 as the input offset current, I_{OS}.

Table 19-3

Resistor	Listed Value	Measured Value
R_1	100 kΩ	
R_2	100 kΩ	

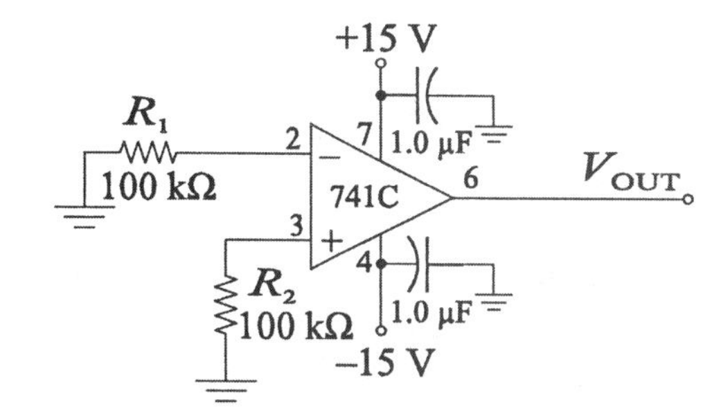

Figure 19-3

4. In this step, you will measure the common-mode rejection ratio, CMRR, of a 741C op-amp. The basic CMRR is the ratio of the op-amp's differential gain ($A_{v(d)}$) divided by the common-mode gain ($A_{v(cm)}$). Because it is a ratio of gains, CMRR is an ac parameter. It is frequently expressed in decibels (indicated with a prime symbol) according to the definition:

$$\text{CMMR}' = 20\log\frac{A_{v\,(d)}}{A_{v\,(cm)}}$$

   (a) Measure and record the resistors listed in Table 19-4. For an accurate measurement of CMRR′, resistors R_A and R_B should be closely matched as should R_C and R_D.
   (b) It is more accurate to compute the differential gain, $A_{v(d)}$, based on the resistance ratio than to measure it directly. Determine the differential gain by dividing the measured value of R_C by R_A. Enter the differential gain, $A_{v(d)}$, in Table 19-1.
   (c) Connect the circuit shown in Figure 19-4. Set the signal generator for 1.0 V_{pp} at 1 kHz. Measure the output voltage, $V_{out(cm)}$. Determine the common-mode gain, $A_{v(cm)}$, by dividing $V_{out(cm)}$ by $V_{in(cm)}$. Record the result in Table 19-1.
   (d) Determine the CMRR′, in decibels, for your 741C op-amp. Record the result in Table 19-1.

Table 19-4

Resistor	Listed Value	Measured Value
R_A	100 Ω	
R_B	100 Ω	
R_C	100 kΩ	
R_D	100 kΩ	

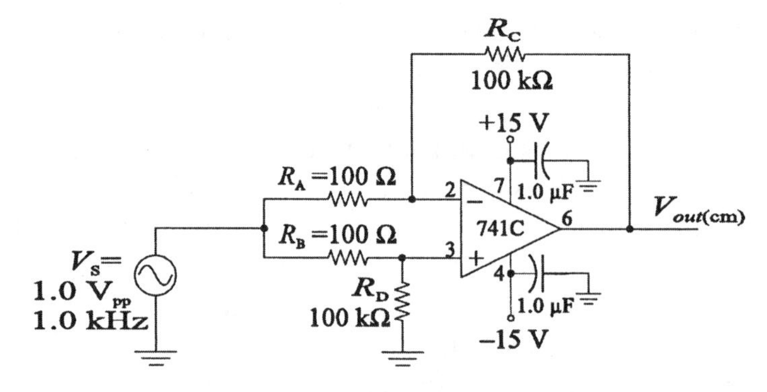

Figure 19-4

5. In this step, you will measure the slew rate of your op-amp. Slew rate is the internally limited rate of change in output voltage with a large-amplitude step function applied to the input. It is usually specified for a unity-gain voltage-follower with a fast rising input pulse. It is usually expressed in units of volts/microsecond (V/µs).

   Connect the unity gain circuit shown in Figure 19-5. Set the signal generator for a 10 V_{pp} square wave at 10 kHz. The output voltage will be slew-rate limited and will not respond instantaneously to the change in the input voltage. The slew rate can be measured by observing the change in voltage divided by the change in time at any two points on the rising output waveform as shown in Figure 19-6. Record the measured value in Table 19-1.

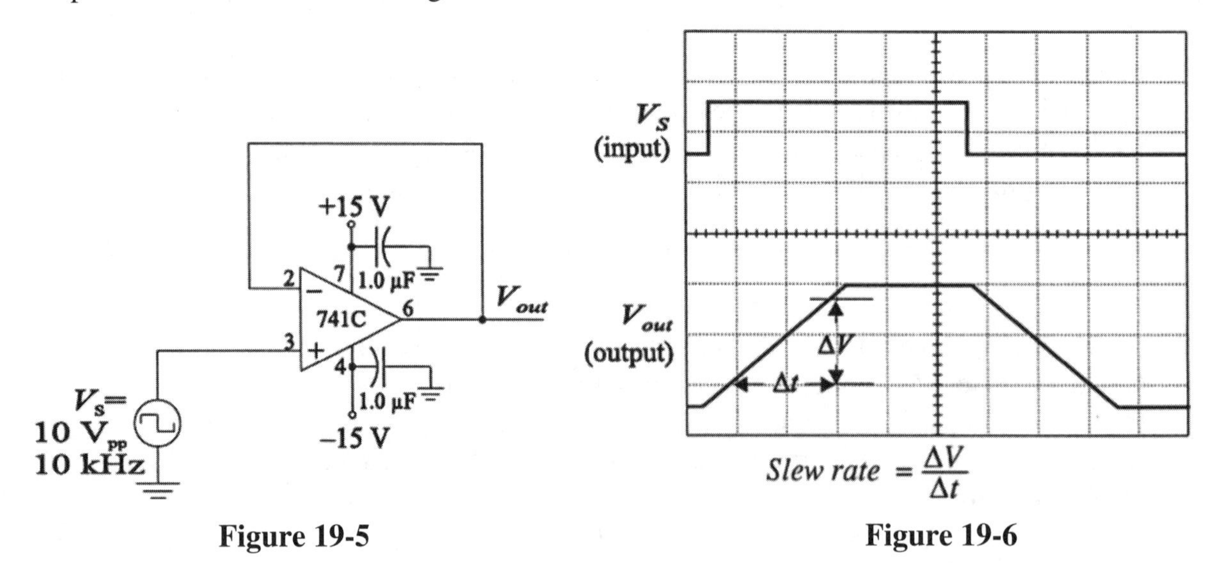

Figure 19-5 **Figure 19-6**

Conclusion:

Evaluation and Review Questions:

1. What is the meaning of the (+) and (–) terminals on the op-amp symbol?

2. Explain the meaning of input offset voltage.

3. What is the difference between the input bias current and the input offset current?

4. (a) What is the difference between differential gain and common-mode gain?

   (b) Explain how you measured the CMRR′ of the 741C.

   (c) What is the advantage of a high CMRR?

5. What is the advantage of a fast slew rate for an op-amp?

For Further Investigation:

Although the output impedance of an op-amp is low, it does not follow that an op amp can drive a very small load impedance. Op-amps have a current-limiting circuit to protect the output when it is short circuited or when the load is too small. Test the current limit of your op-amp using the voltage-follower circuit you constructed in step 5. Reduce the input frequency until you observe no slew rate limiting on the output. While observing, connect a 10 Ω resistor between the output and ground. Sketch the input and output signals and determine the current limit for both the positive and negative excursion of the signal. Compare the measured current limit with the specified current limit. For the signal observed, what is the *smallest* resistor you could use without having current limiting?

Name ______________

Date ______________

Class ______________

20 Linear Op-Amp Circuits

Reading:
Floyd and Buchla, *Analog Fundamentals: A Systems Approach*, Sections 6-4 through 6-6

Objectives:
After performing this experiment, you will be able to:

1. Construct and test inverting and noninverting amplifiers using op-amps.
2. Specify components for inverting and noninverting amplifiers using op-amps.

Summary of Theory:
One of the most important ideas in electronics incorporates the idea of *feedback*, where a portion of the output is returned to the input. If the return signal tends to decrease the input amplitude, it is called *negative feedback*. Negative feedback produces a number of desirable qualities in an amplifier, increasing its stability and its frequency response. It also allows the gain to be controlled independently of the device parameters, temperature, or other variables.

Operational amplifiers are almost always used with external, negative feedback. The feedback circuit determines the specific characteristics of the amplifier. By itself, an op-amp has an extremely high voltage gain called the *open-loop* gain, A_{ol}. When negative feedback is added, the overall gain of the amplifier is determined by the feedback circuit. This gain, including the feedback circuit, is called the *closed-loop* gain, A_{cl}.

Figure 20-1 illustrates a noninverting amplifier with negative feedback. The input is applied to the noninverting terminal and a fraction (B) of the output is returned to the inverting input by the voltage divider. The very high gain of the amplifier forces the two inputs to be very nearly the same voltage; therefore, the input voltage is across R_i and the output voltage is across $R_i + R_f$. The closed-loop gain of this noninverting amplifier is given as $A_{cl(\text{NI})}$. The closed-loop gain is the reciprocal of the feedback fraction as shown with Figure 20-1.

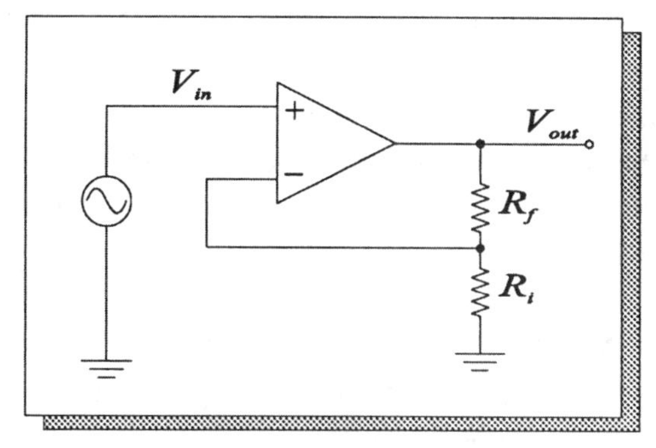

Figure 20-1

$$\text{feedback fraction} = B = \frac{R_i}{R_i + R_f}$$

$$A_{cl\,(\text{NI})} = \frac{1}{B} = \frac{R_i + R_f}{R_i} = 1 + \frac{R_f}{R_i}$$

An inverting amplifier is shown in Figure 20-2. In this amplifier, the output voltage is the opposite phase to the input voltage. The noninverting input is grounded and the input signal is applied through a resistor to the inverting terminal. A feedback resistor is connected between the output and the inverting input. This amplifier can be analyzed by assuming the input current is same as the current in the feedback resistor, $I_{in} = I_f$ and that the open-loop gain of the amplifier is very high. From these assumptions, the closed loop gain can be calculated quite accurately as the ratio of the feedback resistor to the input resistor. The basic equations for the inverting amplifier are shown with Figure 20-2. The minus sign indicates inversion between the input and output. $V_{(-)}$ refers to the inverting input on the op-amp.

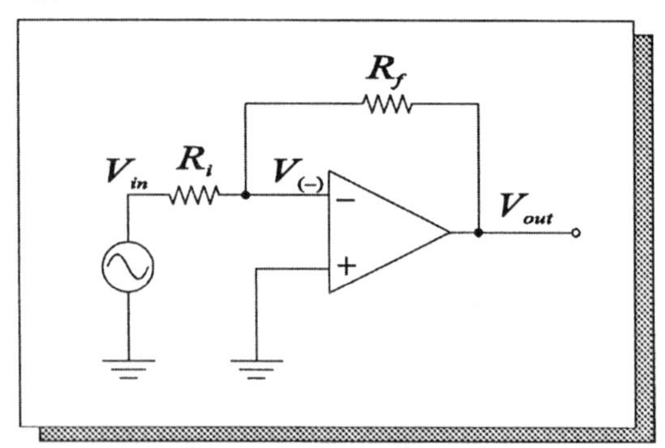

Figure 20-2

$$I_{in} = I_f$$

A_{ol} is very large, therefore

$$V_{(-)} = -\frac{V_{out}}{A_{ol}} \cong 0$$

$$\frac{V_{in}}{R_i} = -\frac{V_{out}}{R_f}$$

$$\frac{V_{out}}{V_{in}} = A_{cl\,(I)} = -\frac{R_f}{R_i}$$

Materials Needed:

Resistors: two 1.0 kΩ, one 10 kΩ, one 470 kΩ, one 1.0 MΩ
Two 1.0 μF capacitors
One 741C op-amp
For Further Investigation:
One 1.0 kΩ potentiometer, one 100 kΩ resistor, assorted resistors to test

Procedure:

1. The circuit to be tested in this step is the noninverting amplifier illustrated in Figure 20-3.
   (a) Measure a 10 kΩ resistor for R_f and a 1.0 kΩ resistor for R_i. Record the measured value of resistance in Table 20-1.
   (b) Using the measured resistances, compute the closed-loop gain of the noninverting amplifier. The closed-loop gain equation is given next to Figure 20-1.
   (c) Calculate V_{out} using the computed closed-loop gain.
   (d) Connect the circuit shown in Figure 20-3. Set the signal generator for a 500 mV$_{pp}$ sinusoidal wave at 1.0 kHz. The generator should have no dc offset.
   (e) Measure the output voltage, V_{out}. Record the measured value.
   (f) Measure the feedback voltage at pin 2. Record the measured value.
   (g) Place a 1.0 MΩ test resistor in series with the generator. Measure the input resistance of the circuit, R_{in}, based on the voltage drop across the test resistor. Use the voltage divider rule to indirectly find the input resistance.

Table 20-1

R_f Measured	R_i Measured	V_{in} Measured	$A_{cl(NI)}$ Computed	V_{out} Computed	V_{out} Measured (pin 6)	$V_{(-)}$ Measured (pin 2)	R_{in} Measured
		500 mV$_{pp}$					

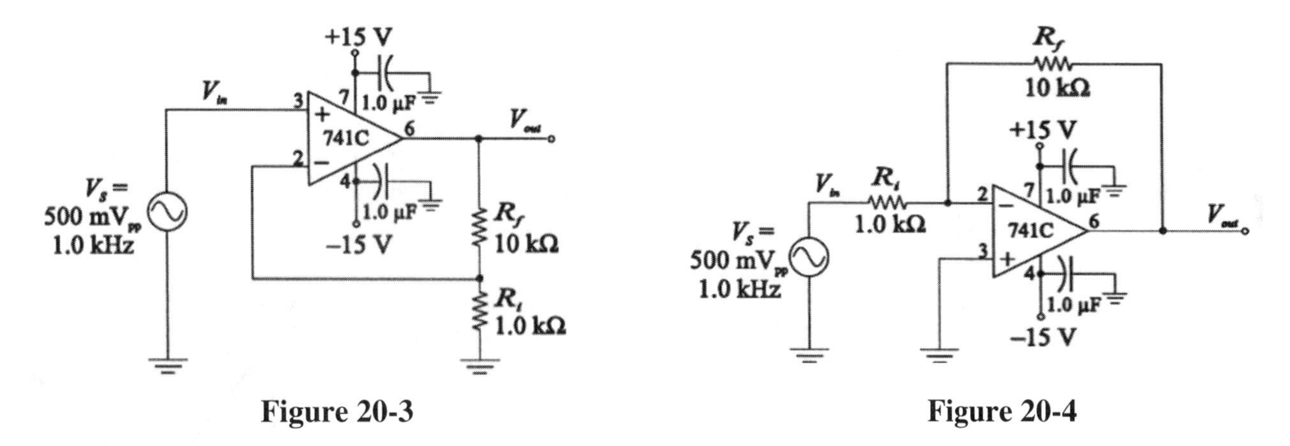

Figure 20-3

Figure 20-4

2. In this step you will test an inverting amplifier. All data are to be recorded in Table 20-2. The circuit is illustrated in Figure 20-4. The closed-loop gain is:

$$A_{cl\ (I)} = -\frac{R_f}{R_i}$$

(a) Use the same resistors for R_f and R_i as in step 1. Record the measured value of resistance in Table 20-2.
(b) Using the measured resistance, compute and record the closed-loop gain of the inverting amplifier.
(c) Calculate V_{out} using the computed closed-loop gain.
(d) Connect the circuit shown in Figure 20-4. Set the signal generator for a 500 mV$_{pp}$ sine wave at 1 kHz. The generator should have no dc offset.
(e) Measure and record the output voltage, V_{out}.
(f) Measure and record the voltage at pin 2. This point is called a *virtual ground* because of the effect of negative feedback.
(g) Place a 1.0 kΩ test resistor in series with the generator and R_i. Measure the input resistance of the circuit, R_{in}, based on the voltage drop across the test resistor. In this case, you should observe that the drop across the test resistor is about the same as the drop across R_i.

Table 20-2

R_f Measured	R_i Measured	V_{in} Measured	$A_{cl(I)}$ Computed	V_{out} Computed	V_{out} Measured (pin 6)	$V_{(-)}$ Measured (pin 2)	R_{in} Measured
		500 mV$_{pp}$					

3. In this step you will specify the components for an inverting amplifier using a 741C op-amp. The amplifier is required to have an input resistance of 10 kΩ and a closed-loop gain of –47. The input test signal is a 1 kHz, 100 mV_{pp} sinusoidal wave signal with no dc component (offset). Check that there is no dc component using an oscilloscope. If necessary, you may need to put a voltage divider on the input to attenuate the signal to the 100 mV level. Draw the amplifier. Then build and test your circuit. *Note:* You need to be careful that the generator does not have a dc offset; remember this is a dc amplifier!

   Find the maximum voltage the input signal can have before clipping occurs. Try increasing the frequency and note the frequency at which the output is distorted. Does the upper frequency response depend on the amplitude of the waveform? Summarize your results in the space provided.

Conclusion:

Evaluation and Review Questions:

1. Express the gain of the amplifiers tested in steps 1 and 2 in dB.

2. It was correct to talk about a *virtual ground* for the inverting amplifier. Why isn't it correct to refer to a virtual ground for the noninverting amplifier?

3. If $R_f = R_i = 10$ kΩ, what gain would you expect for:
   (a) a noninverting amplifier?

   (b) an inverting amplifier?

4. (a) For the noninverting amplifier in Figure 20-3, if $R_f = 0$ and R_i is infinite, what is the gain?

   (b) What is this type of amplifier called?

5. What output would you expect in the inverting amplifier of Figure 20-4 if R_f were open?

For Further Investigation:
An interesting application of an inverting amplifier is to use it as the basis of an ohmmeter for high value resistors. The circuit is shown in Figure 20-5. The unknown resistor, labeled R_x, is placed between the terminals. The output voltage is proportional to the unknown resistance. Calibrate the meter by placing a known 10 kΩ resistor in place of R_x and adjusting the potentiometer for exactly 100 mV output. The output then represents 10 mV/1000 Ω. By reading the output voltage, and moving the decimal point, you can directly read resistors from several thousand ohms to over 1 MΩ.

Construct the circuit and test it using different resistors. Calibrate output voltage against resistance and compare with theory. Find the percent error for a 1 MΩ resistor using a lab meter as a standard. Summarize your results in a short report.

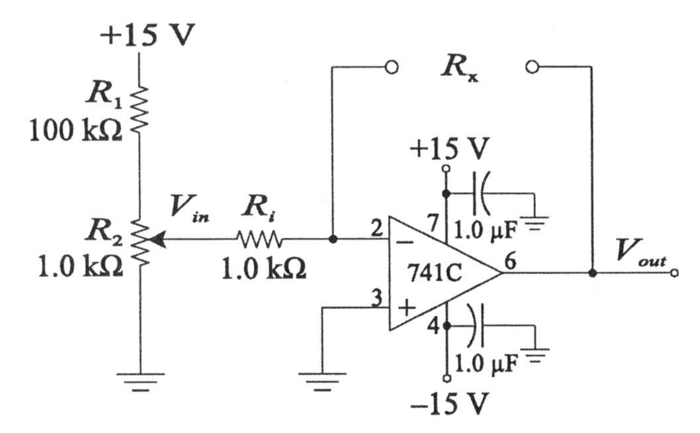

Figure 20-5

Name ________________
Date ________________
Class ________________

21 Op-Amp Frequency Response

Reading:
Floyd and Buchla, *Analog Fundamentals: A Systems Approach*, Sections 7-1 through 7-3

Objectives:
After performing this experiment, you will be able to:
1. Compute and measure the bandwidth of noninverting op-amp circuits as a function of gain.
2. Compute and measure the bandwidth of inverting op-amp circuits as a function of gain.
3. Compare the bandwidth of noninverting op-amp circuits with inverting op-amps circuits.

Summary of Theory:
The typical single-pole frequency response for a 741C op-amp is shown as the curve marked open-loop gain in Figure 21-1. The plot shows the gain plotted as a function of frequency in kHz. The op-amp's cutoff frequency, $f_{c(ol)}$, is the frequency at which the gain is 0.707 of the value at dc. For the 741C, this is typically only about 10 Hz. As with all one-pole *RC* filters, the response rolls off at –20 dB/decade above f_c. This roll-off has a constant slope of –1 (on the log-log plot). This implies that the open-loop gain times the bandwidth is a constant for frequencies above f_c.

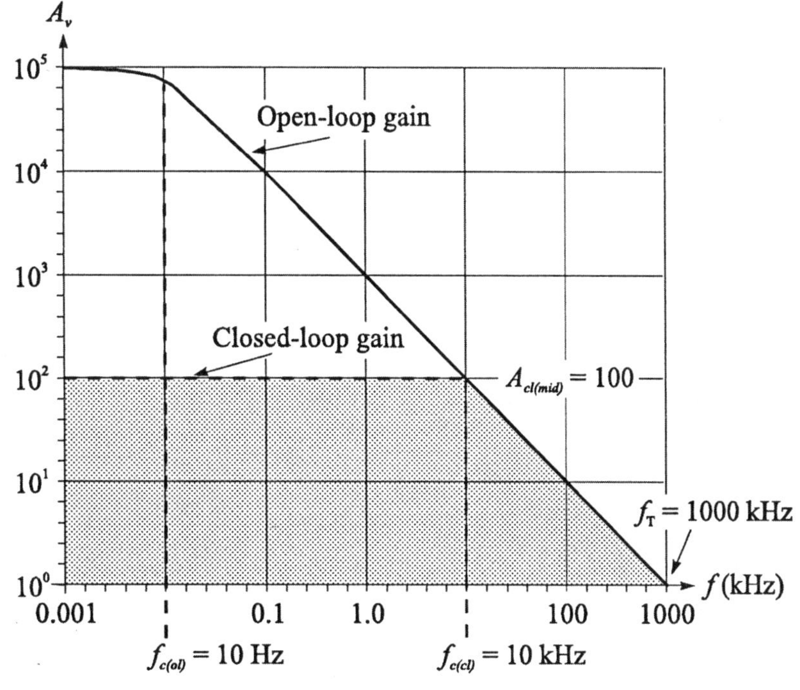

Figure 21-1

Closed-loop refers to the circuit with negative feedback present to control the gain and other parameters. When an op-amp is configured as a noninverting amplifier, the closed-loop gain is $1/B$ as discussed in Experiment 20 (recall that B is the feedback fraction). If the closed-loop gain of a noninverting amplifier is 1.0, the bandwidth is given by f_T, the unity-gain frequency (see Figure 21-1). For other gains, the closed-loop bandwidth (equivalent to the cutoff frequency for any dc amplifier) can be determined graphically from the frequency response curve by noting the intersection of the closed-loop gain line with the open-loop response. An example is shown in Figure 21-1. For a closed-loop gain of 100 (shaded area) the bandwidth is observed to be 10 kHz. The same result can be obtained from the equation

$$BW_{cl} = f_T B \qquad \textit{Equation 21-1}$$

where:

BW_{cl} = closed-loop bandwidth
f_T = unity-gain frequency
B = feedback fraction

The bandwidth for an inverting amplifier is not as high as a comparable noninverting amplifier. The difference is small for high-gain circuits but is more pronounced for lower gain circuits or with certain applications such as a summing amplifier. To compute the bandwidth for the inverting configuration, assume all signal sources are at ground and calculate the feedback fraction, B, *as if the noninverting input were driven.* Use this B in Equation 21-1 to find the bandwidth.

For example, consider the inverting amplifier in Figure 21-2(a) with a unity-gain frequency of 1.0 MHz and a gain of –2.0. To compute the bandwidth, mentally move the input to the noninverting terminal as in (b) and compute the feedback fraction. In this case, $B = 1/3$. Substituting into Equation 21-1,

$$BW_{cl} = f_T B$$
$$= (1.0 \text{ MHz})(1/3) = 333 \text{ kHz}$$

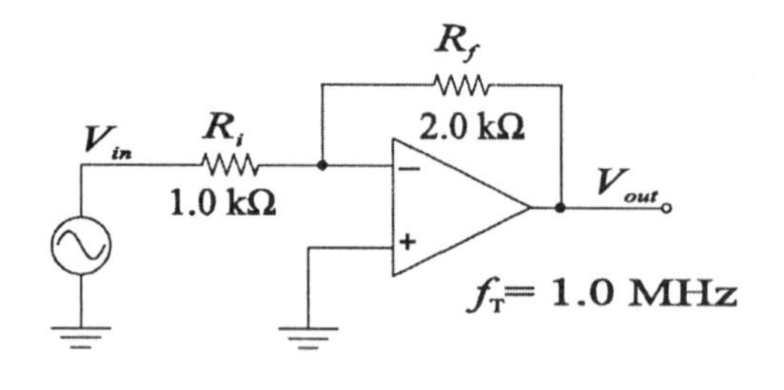

(a) Inverting amplifier with a gain of –2

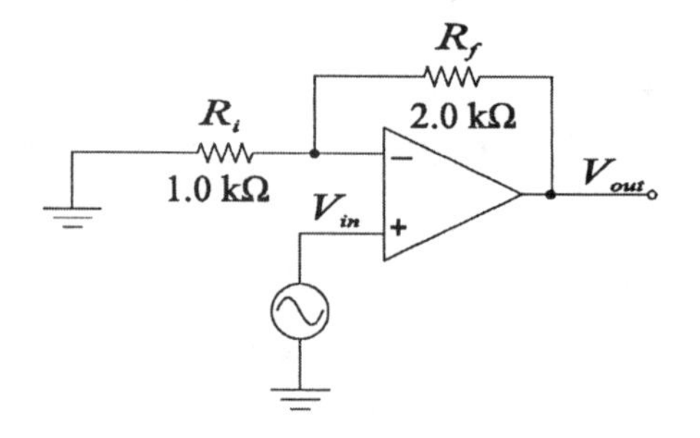

(b) Calculate frequency response based on driving noninverting input

Figure 21-2

Materials Needed:

Resistors: one 620 Ω, two 1.0 kΩ, one 2.0 kΩ, one 3.3 kΩ, one 10 kΩ, one 18 kΩ, one 100 kΩ
Two 1.0 μF capacitors
One 741C op-amp

For Further Investigation:

One 100 Ω resistor

Procedure:

Frequency Response of the Noninverting Amplifier

1. Measure and record the values of the resistors listed in Table 21-1.

2. Construct a noninverting voltage-follower as in Figure 21-3 with the power supplies set to ±15 V. Connect 1.0 μF bypass capacitors near the power connections to the op-amp. The purpose of R_A and R_B is to attenuate the generator voltage; they are not necessary if you can set your generator to a 50 mV_{pp} sine wave.

Table 21-1

Step	Resistor	Listed Value	Measured Value
4	$R_{i\text{-}1}$	2.0 kΩ	
4	$R_{f\text{-}1}$	18 kΩ	
5	$R_{i\text{-}2}$	1.0 kΩ	
5	$R_{f\text{-}2}$	100 kΩ	

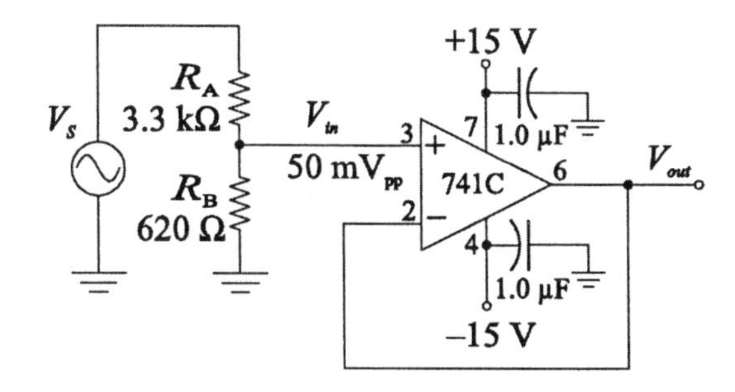

Figure 21-3

3. (a) To measure the unity-gain frequency (or unity-gain bandwidth), a very small signal must be used to avoid slew rate limitations. Adjust the generator for a 50 mV_{pp} sine wave at the noninverting input terminal of the op-amp at a frequency of 10 kHz. Measure the output voltage and record the gain in Table 21-2.

   (b) Increase the frequency until the output amplitude falls to 70.7% of the output amplitude observed for the gain measurement. Adjust the generator as necessary to maintain the input signal at 50 mV_{pp}. Since this is a dc amplifier, the closed-loop bandwidth is equal to the unity-gain frequency. Measure and record the bandwidth in Table 21-2.

Table 21-2 Data for Noninverting Amplifiers

Step	Computed Gain	Measured Gain	Closed-loop bandwidth Computed	Closed-loop bandwidth Measured
3	1.0			
4	10			
5	101			

4. (a) Change the circuit to a noninverting amplifier with a gain of 10 as shown in Figure 21-4. Set the frequency to 10 kHz with a 50 mV_{pp} input signal. Measure the output voltage and compute the gain. Enter the measured gain in Table 21-2.

   (b) Compute the closed-loop bandwidth for this configuration by dividing the unity-gain bandwidth (step 3) by the measured gain (step 4(a)). Now measure the bandwidth by increasing the generator's frequency until the output amplitude falls to 70.7% of the output amplitude that was measured at 10 kHz. Again, the input signal should be maintained at 50 mV_{pp}. Record the data in Table 21-2.

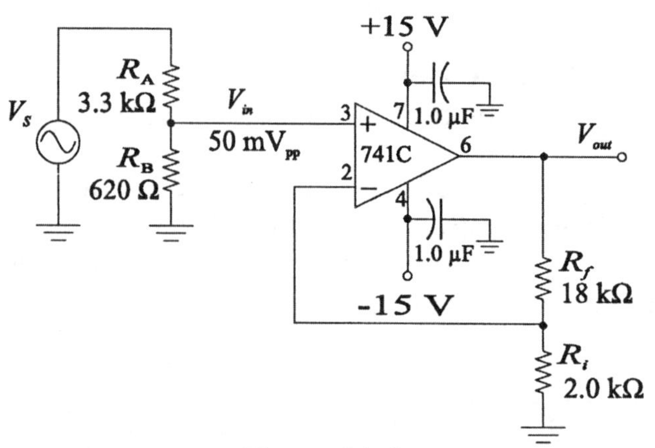

Figure 21-4

5. Change R_f to 100 kΩ and R_i to 1.0 kΩ, resulting in a computed gain of 101. Repeat step 4 but start the frequency at 1.0 kHz with a 50 mV$_{pp}$ input signal. Measure the gain and the bandwidth as before. Record the values in Table 21-2.

Frequency Response of the Inverting Amplifier

6. Measure and record the values of the resistors listed in Table 21-3.

Table 21-3

Step	Resistor	Listed Value	Measured Value
7	$R_{i\text{-}3}$	1.0 kΩ	
7	$R_{f\text{-}3}$	1.0 kΩ	
8	$R_{f\text{-}4}$	10 kΩ	
9	$R_{f\text{-}5}$	100 kΩ	

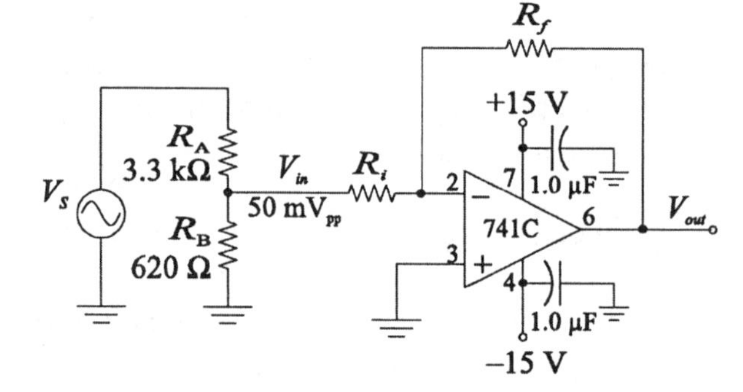

Figure 21-5

7. Refer to Figure 21-5. Using 1.0 kΩ resistors for both ($R_{i\text{-}3}$ and $R_{f\text{-}3}$), construct an inverting amplifier with a gain of –1.0. Compute the closed-loop bandwidth using Equation 21-1. Assume the unity-gain frequency is the same as you found in step 3 and the feedback fraction is 0.5 (why?). Measure the gain and bandwidth as before using a 50 mV$_{pp}$ signal as the input. Record computed and measured values in Table 21-4.

8. Form an inverting amplifier with a gain of –10 by changing R_f equal to 10 kΩ (Use $R_{f\text{-}4}$). Compute the bandwidth for this configuration. Repeat the measurements of gain and bandwidth, and record these values in Table 21-4.

Table 21-4 Data for Inverting Amplifiers

Step	Computed Gain	Measured Gain	Closed-loop bandwidth Computed	Closed-loop bandwidth Measured
7	–1.0			
8	–10			
9	–100			

9. Form an inverting amplifier with a gain of –100. (Use $R_{f\text{-}5}$). Compute the bandwidth for this configuration. Repeat step 7 but start the frequency at 1.0 kHz with a 50 mV_{pp} input signal. Record the data in Table 21-4.

Conclusion:

Evaluation and Review Questions:

1. (a) From your data for the noninverting amplifier (Table 21-2), determine if the gain-bandwidth product was constant.

   (b) Should it be?

2. (a) From your data for the inverting amplifier (Table 21-4), determine if the gain-bandwidth product was constant.

   (b) Should it be?

3. In step 5, you were directed to start the gain and bandwidth measurements using a 1.0 kHz signal instead of a 10 kHz signal as before. Explain why this was necessary.

4. (a) How would you expect the bandwidth of the amplifiers in this experiment to affect the rise time of a square wave input?

   (b) What factor, other than bandwidth, can also affect the rise time of a square wave?

5. Is it possible to increase the bandwidth of a high-gain amplifier by substituting two lower gain amplifiers? Explain.

For Further Investigation:
The bandwidth can be controlled for an inverting amplifier by connecting a resistor from the inverting input to ground. Investigate this effect by testing the circuit shown in Figure 21-6. The circuit is essentially the same as Figure 21-5 but with the addition of R_C. Find the gain at a low frequency; then find the bandwidth. Consider the feedback fraction as seen from the noninverting terminal. Compare your results with those in Table 21-4. Can you explain the significance of the reduced bandwidth? Can you think of an application where a reduced bandwidth like this is desirable?

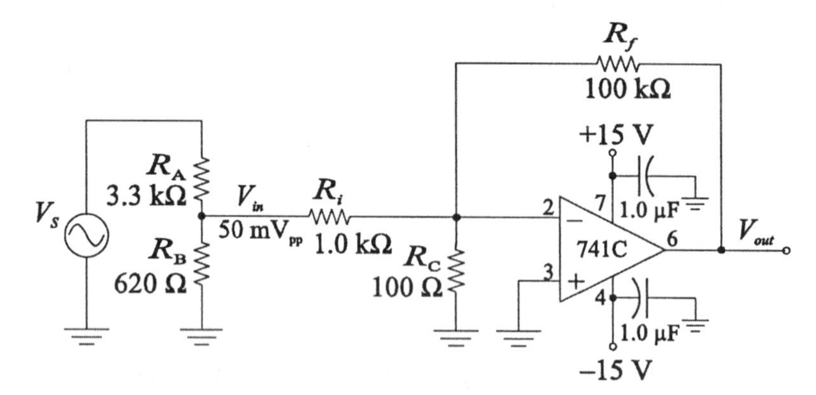

Figure 21-6

Name ____________
Date ____________
Class ____________

22 Comparators and the Schmitt Trigger

Reading:
Floyd and Buchla, *Analog Fundamentals: A Systems Approach*, Section 8-1

Objectives:
After performing this experiment, you will be able to:
1. Compare the input and output waveforms for comparator and Schmitt trigger circuits.
2. Use an oscilloscope to plot the transfer curve for a comparator circuit, including one with hysteresis.
3. Construct and test a relaxation oscillator using a Schmitt trigger.

Summary of Theory:
A comparator is a switching device that produces a high or low output, depending on which of the two inputs is larger. The comparator is run with the very high open-loop gain. When the noninverting input is very slightly larger than the inverting input, the output goes to positive saturation; otherwise it goes to negative saturation. Although general-purpose op-amps can be used as comparators, specially designed op-amps can switch faster and have additional features not found on general-purpose op-amps. For noncritical applications, general-purpose op-amps are satisfactory and will be used in this experiment.

A comparator circuit is characterized by its transfer characteristic. The transfer characteristic (curve) is a plot of the output voltage (plotted along the *y*-axis) as a function of the input voltage (plotted along the *x*-axis). Consider the comparator shown in Figure 22-1(a). The reference voltage is +6 V. When the input is greater than +6.0 V, the output will go to positive saturation (approximately +13 V); when the input is less than +6.0 V, the output will be in negative saturation. A glance at the transfer curve will show the output for any given input voltage.

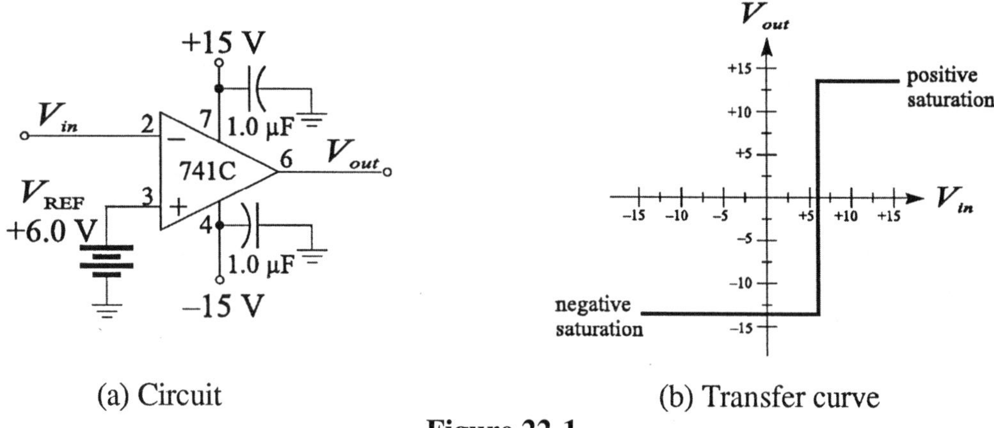

(a) Circuit

(b) Transfer curve

Figure 22-1

Because of the sensitivity to a small input change, the output of a comparator may change due to noise on the input or when the input changes slowly. To avoid this, hysteresis is added to the comparator circuit by introducing positive feedback. The circuit, called a *Schmitt trigger*, has two switching thresholds – one for a rising input voltage, the other for a falling input. By separating the two thresholds, noise effects can be eliminated. In this experiment, you will investigate two comparators and an inverting Schmitt trigger circuit.

Materials Needed:
Resistors: one 100 kΩ
Two 1.0 μF capacitors
One 10 kΩ potentiometer
One 741C op-amp
For Further Investigation:
One additional 100 kΩ resistor
One 0.1 μF capacitor

Procedure:
Comparator and Comparator Transfer Curve

1. Figure 22-2 shows an inverting comparator circuit with a variable threshold determined by the potentiometer setting. Construct the circuit and set V_{REF} to near 0 V. Set the function generator for a 3.0 V_{pp} triangle waveform at 50 Hz and observe the input and output waveforms on a two-channel oscilloscope. Sketch the waveforms on Plot 22-1. Note the point where switching takes place. Be sure to label all plots in this experiment.

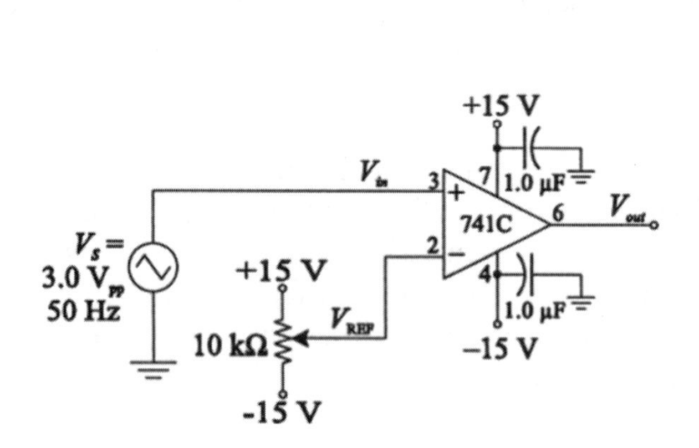

Figure 22-2

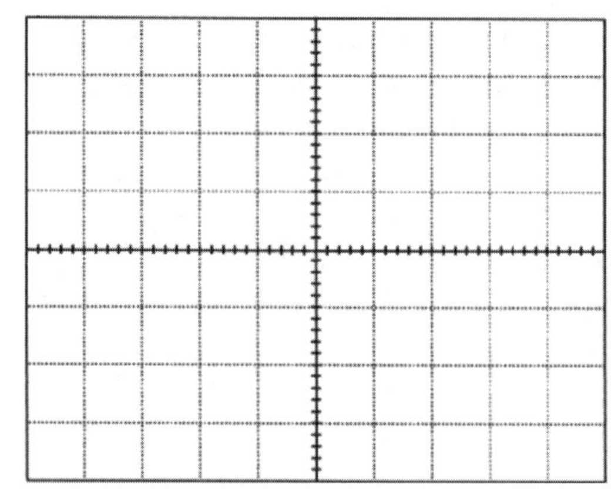

Plot 22-1 Comparator waveform.

2. Observe the output as you vary the potentiometer. Then reset V_{REF} to 0 V.

   Observations:____

3. In this step, you will plot the transfer curve for the comparator on the oscilloscope. Place V_{in} on the X-channel and V_{out} on the Y-channel. Set the VOLTS/DIV control so that both signals are on the screen. Neither channel should be inverted. Then switch the oscilloscope to the X-Y mode. Sketch (and label) the transfer curve you see in Plot 22-2.[1]

[1] You may observe a slight difference in the switching point, depending on whether the input rises or falls, due to the

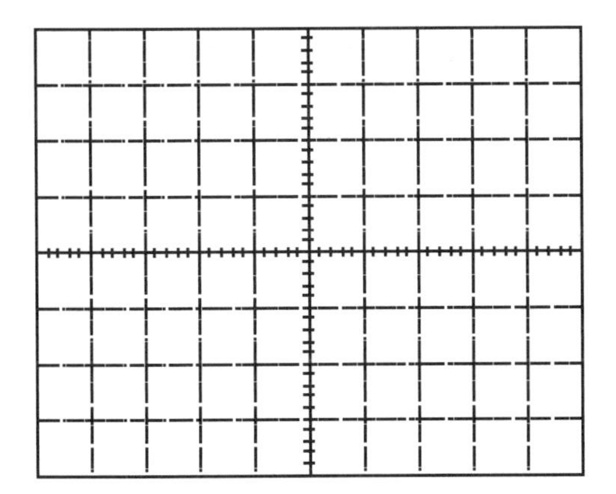

Plot 22-2 Comparator transfer curve.

Plot 22-3 Comparator transfer curve (inputs reversed).

4. Vary the potentiometer as you observe the transfer curve.

   Observations:____

5. While observing the transfer curve, reverse the inputs to the comparator. Sketch the new transfer curve in Plot 22-3.

Schmitt Trigger and Schmitt Trigger Transfer Curve

6. Construct the Schmitt trigger circuit shown in Figure 22-3. Set the potentiometer to the maximum resistance and put the oscilloscope in normal time base mode (not X-Y mode). Slowly reduce the resistance of the potentiometer and observe the input and output waveforms. Note that when the output changes states, the input voltage is different for a rising and a falling signal. Sketch the observed input and output waveforms in Plot 22-4.

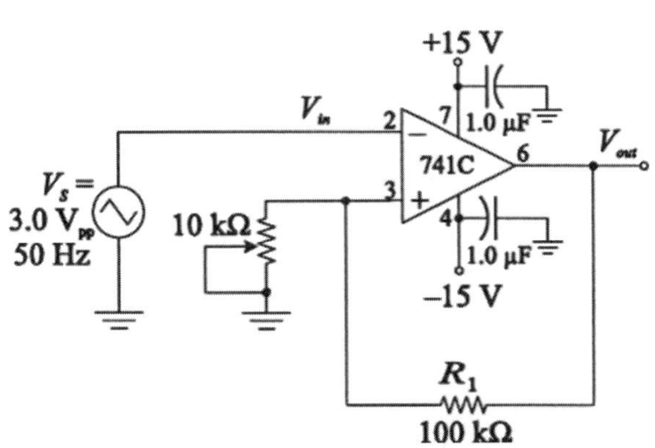

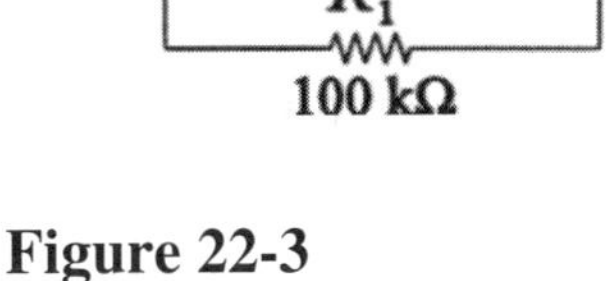

Figure 22-3

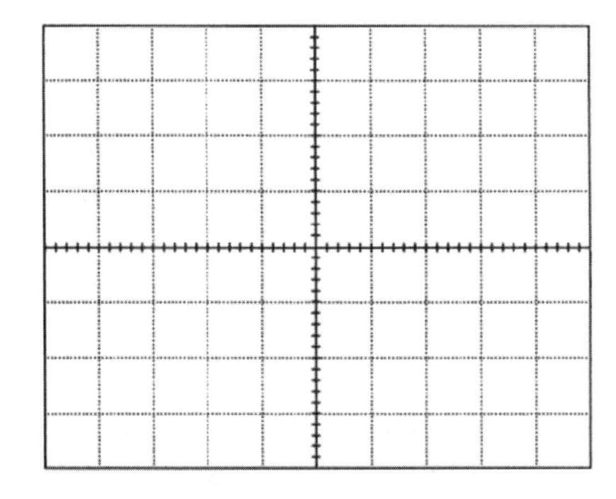

Plot 22-4 Schmitt trigger waveform.

---

slew rate limitation of the op-amp. Slowing the generator will reduce the effect.

7. In this step, you will plot the transfer curve for the Schmitt trigger on the oscilloscope. The input signal is again on the X-channel and the output signal is on the Y-channel. Select the X-Y mode and adjust the controls to view the transfer curve. Notice the hysteresis. Sketch (and label) the transfer curve you see in Plot 22-5.

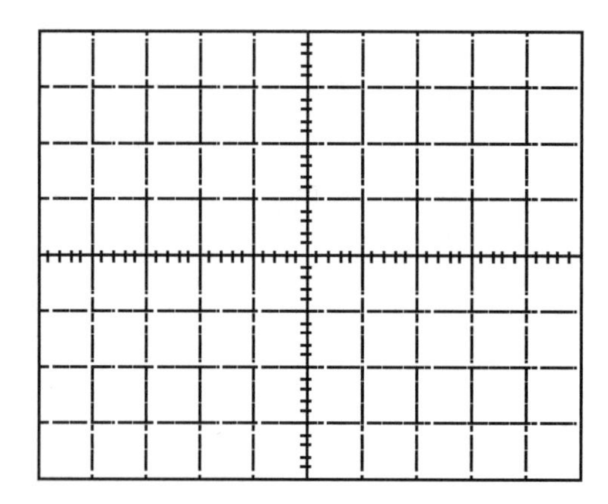

Plot 22-5 Schmitt trigger transfer curve.

8. While observing the transfer curve, vary the potentiometer.

   Observations:____

Conclusion:

Evaluation and Review Questions:

1. Describe how the threshold voltage changes the transfer curve for a comparator.

2. Assume the circuit in Figure 22-2 had V_{REF} set to zero volts. How would you expect the output to be affected by varying the dc offset control on the generator?

3. Would a sinusoidal input to the comparators produce the same transfer curve as a triangle waveform? Explain.

4. Summarize the important differences between a comparator and a Schmitt trigger.

5. Assume the input signal in Figure 22-3 could have as much as 100 mV_{pp} noise. In order to avoid multiple tripping due to noise, you need to set the trip points at least 100 mV apart. What is the minimum value of resistance that the potentiometer can be set? Assume the output saturates at ±13 V.

For Further Investigation:

The Schmitt trigger you investigated in the experiment can be modified slightly to form a relaxation oscillator. Investigate the relaxation oscillator shown in Figure 22-4. (For simplicity, power supply connections are not shown.) Look at the waveform across the capacitor and at the output. Test the effect of the potentiometer on the frequency and waveform on the capacitor. Normally, a triangle waveform is observed. Under what conditions can you obtain a sine wave? Then observe the transfer curve by observing the signal on pin 2 and pin 6 in the X-Y mode – can you explain the shape? Summarize your findings in a short report.

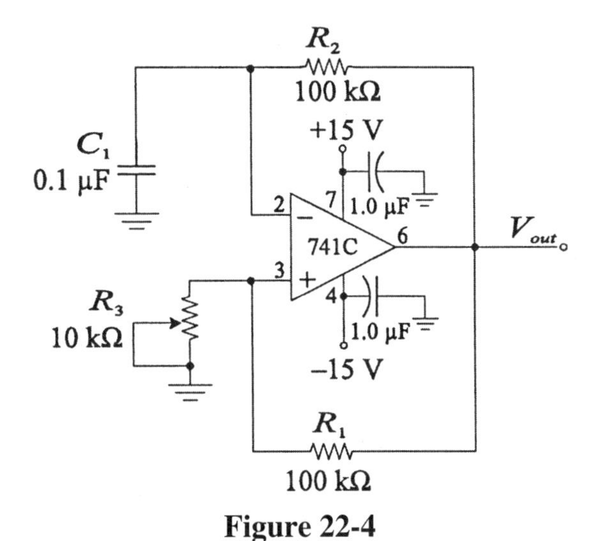

Figure 22-4

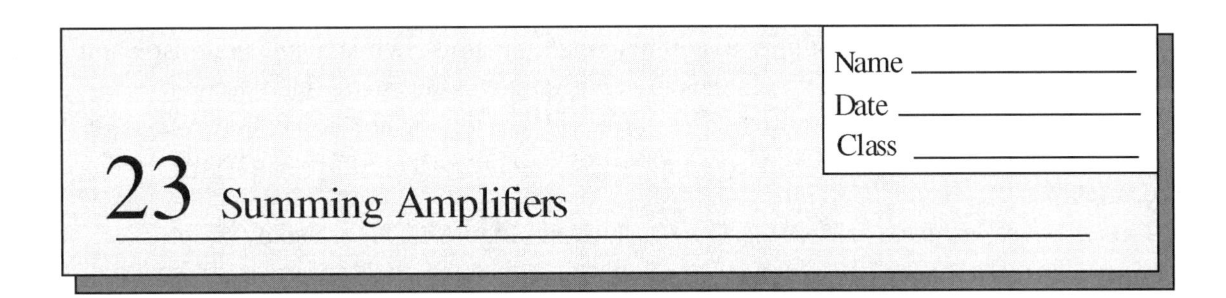

23 Summing Amplifiers

Reading:
Floyd and Buchla, *Analog Fundamentals: A Systems Approach*, Section 8-2

Objectives:
After performing this experiment, you will be able to:
1. Construct and test a digital-to-analog converter (DAC) from a summing amplifier. Apply a binary count sequence to form a step generator, and explain the waveforms.
2. Test various precision diode circuits including a noninverting half-wave rectifier, an inverting half-wave rectifier, and a full-wave rectifier that uses a summing amplifier.
3. Explain the operation of these circuits.

Summary of Theory:
The summing amplifier, illustrated in Figure 23-1, is a multiple input version of an inverting amplifier. The current into the feedback resistor, R_f, is the sum of the current in each input resistor. Because the inverting input is at virtual ground, the total input current is $V_1/R_1 + V_2/R_2 + V_3/R_3$. The virtual ground has the advantage of isolating the various inputs from each other. In addition, the gain of each input can be set differently. Variable gain is useful for mixing several audio sources at different levels (such as microphones) into a single channel.

Another application of a summing amplifier is the notch or band-reject filter illustrated in Figure 23-2. In this circuit, the outputs of a low-pass and a high-pass filter are combined in a summing amplifier. Typically, the low-pass and high-pass filters are designed with operational amplifiers to form the filter elements (this defines an *active filter*). Designs based on this idea are called state-variable filters and they are the most versatile type of active filter. Outputs could be taken from both the low-pass and high-pass filter sections in addition to the output of the summing amplifier. (Active filters are studied in Chapter 9 of the text.)

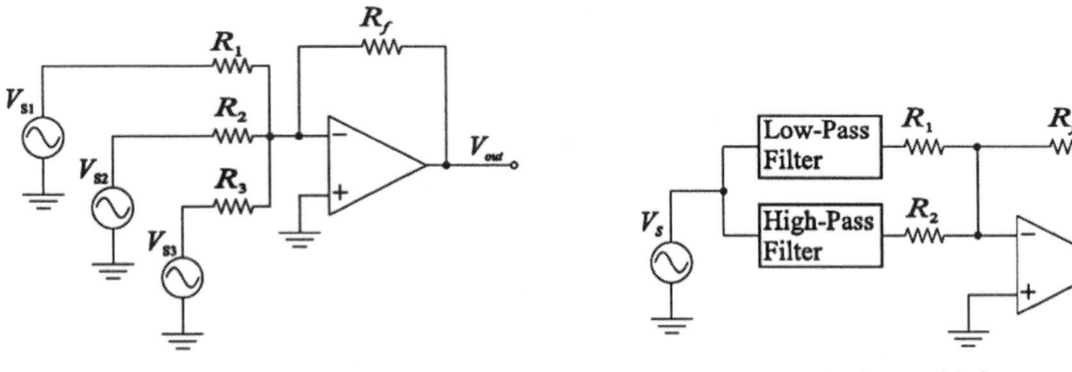

Figure 23-1

Figure 23-2

The ability of a summing amplifier to combine multiple inputs while maintaining isolation between them leads to a number of other interesting applications. One circuit that you will investigate is a precision full-wave rectifier. (Precision refers to the fact that the output is not offset by any diode drops.) The full-wave circuit contains a precision inverting half-wave rectifier and a summing amplifier.

A *noninverting* precision half-wave circuit is shown in Figure 23-3, formed by simply inserting a diode into the feedback path of a voltage-follower. When the diode is forward-biased, it closes the feedback loop, giving an output that follows the input exactly. When the diode is reverse-biased, the output essentially disconnects the op-amp from the load, causing the output to be zero.

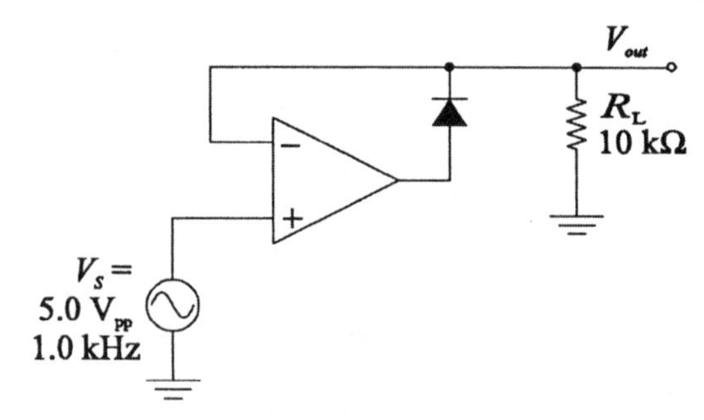

Figure 23-3 Noninverting precision half-wave rectifier.

A summing amplifier can also be used as a basic digital-to-analog converter (DAC) by weighting the inputs according to a binary sequence. Adding a counter to the input results in a step-generator that produces a series of steps by the weighted addition of the outputs of the counter for applications such as a transistor curve tracer. This experiment starts with this circuit, as an interface between the digital and analog world.

Materials Needed:
Resistors: one 3.9 kΩ, one 5.1 kΩ, four 10 kΩ, one 20 kΩ
Capacitors: two 1.0 μF
Two signal diodes, 1N914 (or equivalent)
Two op-amps: LM741C
One 7493A 4-bit ripple counter
For Further Investigation:
Resistors: two 4.7 kΩ, three 100 kΩ
Capacitors: one 0.01 μF, one 0.1 μF

Procedure:
DAC and Step Generator

1. Measure and record the values of the resistors listed in Table 23-1.

2. The circuit shown in Figure 23-4 is a summing amplifier connected to the outputs of a binary counter. The counter outputs are weighted differently by resistors R_A through R_C, and added by the summing amplifier. The resistors and summing amplifier form a basic DAC. Note that the 7493A counter is powered from a +5.0 V supply. The input to the 7493A is a logic pulse (approximately 0 to 4 V) at 1.0 kHz from a function generator. Construct the circuit. Observe V_{out} from the 741C. You should observe a series of steps. Sketch the output in Plot 23-1. Label the voltage and time on your plot.

Table 23-1

Resistor	Listed Value	Measured Value
R_A	20 kΩ	
R_B	10 kΩ	
R_C	5.1 kΩ	
R_f	3.9 kΩ	

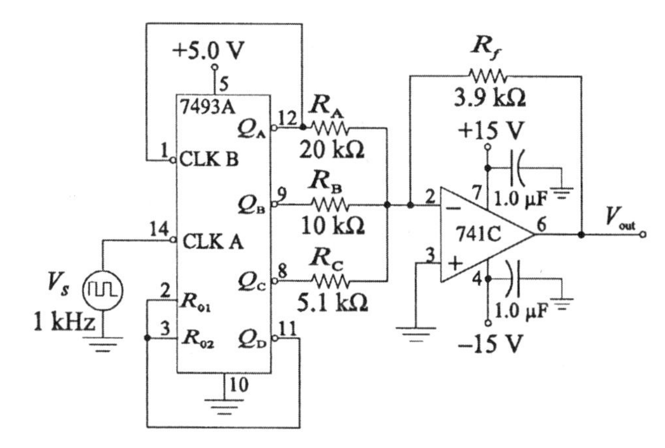

Figure 23-4

3. To see how the steps are formed, observe the Q_A, Q_B, and Q_C outputs from the 7493A. To see the correct time relationship between the signals, put Q_C on channel 1 of your oscilloscope; trigger the scope from this channel. Keep channel 1 in place while moving the channel 2 probe. Sketch the waveforms in the correct time relation in Plot 23-2.

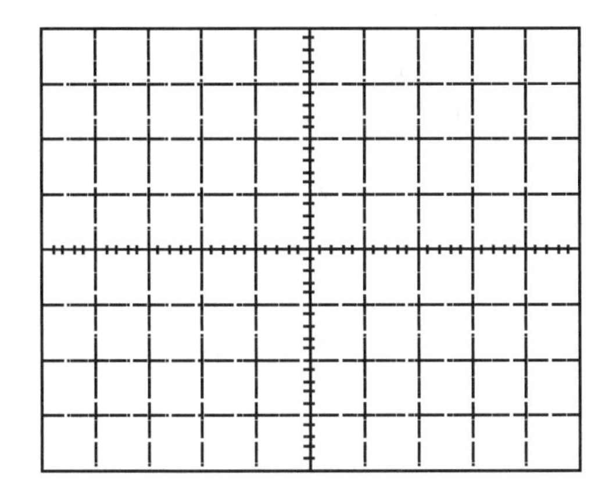

Plot 23-1

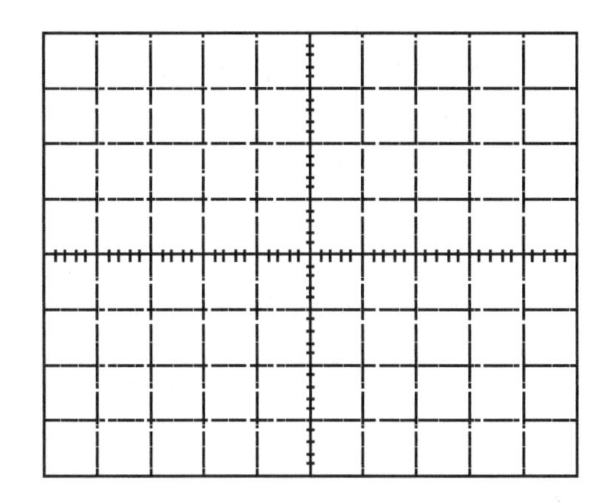

Plot 23-2

Precision Noninverting Half-wave Rectifier

4. A precision noninverting half-wave rectifier was shown in Figure 23-3 and discussed in the Summary of Theory. Construct the circuit; set the input waveform for a 5.0 V_{pp} sinusoidal wave with no dc offset at a frequency of 1.0 kHz. The power connections on this and remaining circuits in this experiment are not shown explicitly – connect the 741C the same as before including the bypass capacitors. Observe the output waveform. You should see that the output follows the input almost exactly except for a small "jump" on the leading edge. (This jump is more pronounced if you raise the frequency). The reason for the jump is the time required (slew rate) for the output to go from negative saturation to +0.7 V (the voltage required to turn on the diode).

Observations:____

____

Precision Inverting Half-wave Rectifier

5. A precision inverting half-wave rectifier is shown in Figure 23-5. The diode between the op-amp output and the inverting input (D_1) prevents the output from saturating, allowing the output to change immediately after the diode starts conducting. The circuit can be recognized as an inverting amplifier with a diode added to the feedback path and the clamping diode. Construct the circuit with the input set as before and observe the output. Look at pin 6 and momentarily pull D_1 from the circuit.

   Observations:____

   ____

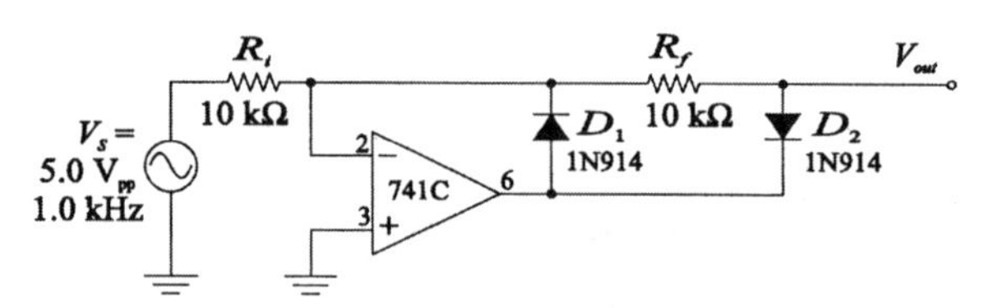

Figure 23-5

Precision Full-wave Rectifier

6. By combining the inverting half-wave rectifier from step 5 with a summing amplifier, a precision full-wave rectifier can be constructed. The circuit is shown in Figure 23-6. Construct the circuit with the input set as before. On Plot 23-3, sketch the waveforms at the left side of R_{i2} and R_{i3} (inputs) and V_{out}.

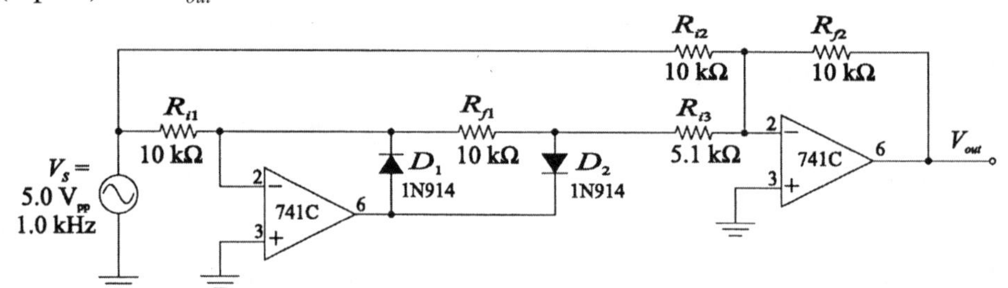

Figure 23-6

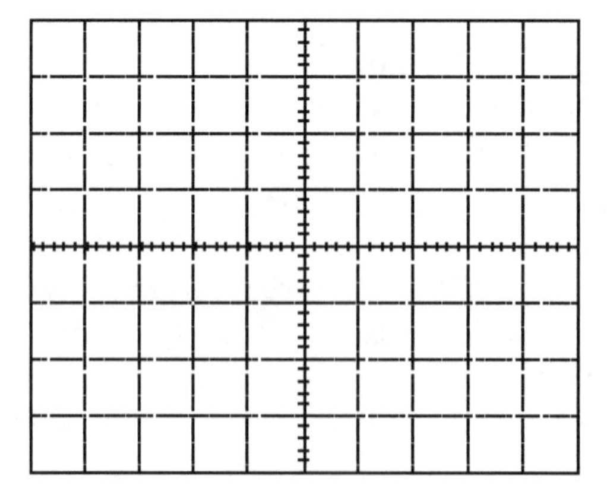

Plot 23-3

Conclusion:

Evaluation and Review Questions:

1. (a) The step generator in Figure 23-4 forms negative falling steps starting at zero volts and going to a negative voltage (approximately –4.4 V). Explain why.

   (b) How could you modify the circuit to produce positive, rising steps at the output?

2. Assume that all three inputs to the summing amplifier (Q_A, Q_B, and Q_C) in Figure 23-4 are 4.5 V. Compute the output voltage from the summing amplifier.

3. Assume you have a function generator that does not have a dc offset control. Show how you could use a summing amplifier to add or subtract a dc offset from the output.

4. The gain for the summing amplifier in the full-wave rectifier circuit (Figure 23-6) is not the same for both inputs. Explain why.

5. The word *operational amplifier* originated from mathematical operations that could be performed with it. Assume you wanted to produce a circuit for which the output voltage was given by the expression $V_{out} = -3A - 2B$ (A and B are variable input voltages). Show how this operation could be accomplished with a summing amplifier by drawing the circuit. Show values for resistors.

For Further Investigation:

As discussed in the Summary of Theory, a notch filter can be constructed by summing the outputs of a low-pass and a high-pass filter in a summing amplifier. Designs based on this idea are called state-variable filters and they are the most versatile of active filters. The particular filter shown in Figure 23-7 replaces the normal active filters with passive ones to give you an idea of the process. Start with the generator set to a 300 mV_{pp} sine wave at 100 Hz. Then investigate the response of the circuit by varying the frequency across the audio band while you monitor the output on an oscilloscope. Summarize your findings.

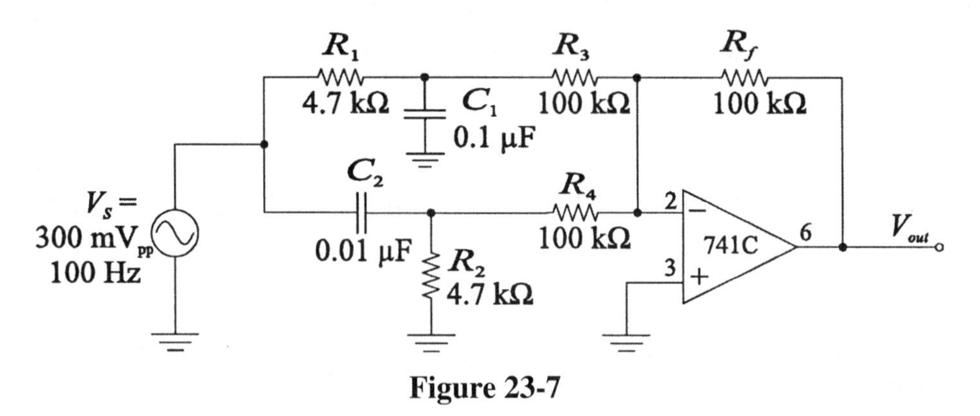

Figure 23-7

Name ______________
Date ______________
Class ______________

24 The Integrator and Differentiator

Reading:
Floyd and Buchla, *Analog Fundamentals: A Systems Approach*, Section 8-3

Objectives:
After performing this experiment, you will be able to:
1. Construct and test integrator and differentiator circuits.
2. Determine the response of the circuits listed in objective 1 to various waveforms.
3. Troubleshoot faults in integrator and differentiator circuits.

Summary of Theory:
A nonlinear application of an op-amp is the comparator, introduced in Experiment 22 to form a square wave (and again in the For Further Investigation to make variable duty cycle pulses). You will also investigate two other nonlinear circuits that have application in waveform generation and signal processing – the integrator and the differentiator. An integrator produces an output voltage that is proportional to the *integral* (sum) of the input voltage waveform over time. The opposite of integration is differentiation. Differentiation means finding the rate of change. A differentiator circuit produces an output that is proportional to the *derivative* or rate of change of the input voltage over time. Basic op-amp integrator and differentiator circuits with a square wave input are illustrated in Figure 24-1.

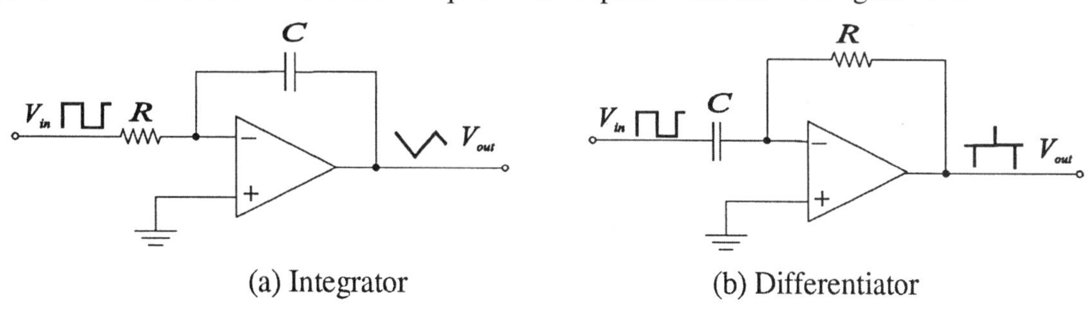

(a) Integrator (b) Differentiator

Figure 24-1

Materials Needed:
Resistors: two 1.0 kΩ, four 10 kΩ, two 22 kΩ, one 330 kΩ
Capacitors: one 2200 pF, one 0.01 μF, two 1.0 μF
Three 741C op-amps
One 1.0 kΩ potentiometer
Two LEDs (one red, one green)

For Further Investigation:
Resistors: two 4.7 kΩ, two 100 kΩ
One 0.1 μF capacitor

Procedure:

1. Construct the comparator circuit shown in Figure 24-2. Vary the potentiometer. Measure the output voltage when the red LED is on and then when the green LED is on. Record the output voltages, V_{OUT}, in Table 24-1. Then set the potentiometer to the threshold point. Measure and record V_{REF} at the threshold. It should be very close to 0 V.

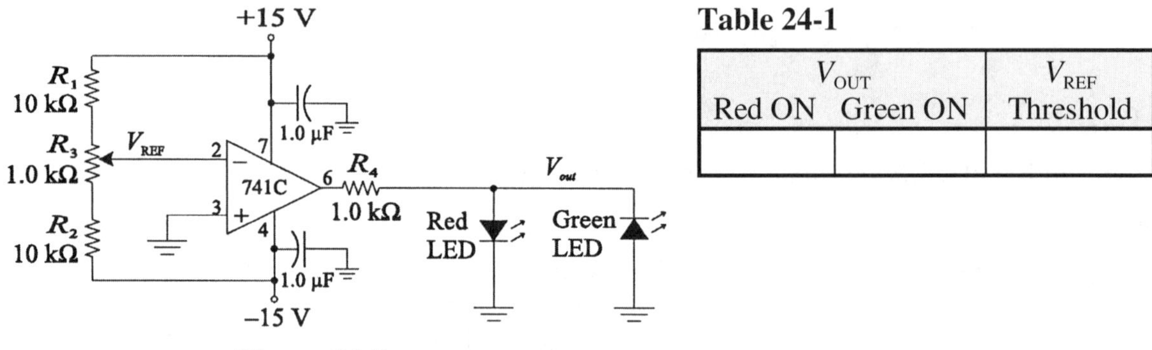

Figure 24-2

Table 24-1

V_{OUT} Red ON	V_{OUT} Green ON	V_{REF} Threshold

2. In this step, you will test the effects of the comparator on a sinusoidal wave input and add an integrating circuit to the output of the comparator. Connect the circuit shown in Figure 24-3 with a 1.0 V_{pp} sine wave input at 1.0 kHz as illustrated. Check that there is no dc offset. Observe the waveforms from the comparator (point **A**) and from the integrator (point **B**). Adjust R_3 so that the waveform at **B** is centered about zero volts. Sketch the observed waveforms in the correct time relationship on Plot 24-1. Show the voltages and time on your plot.

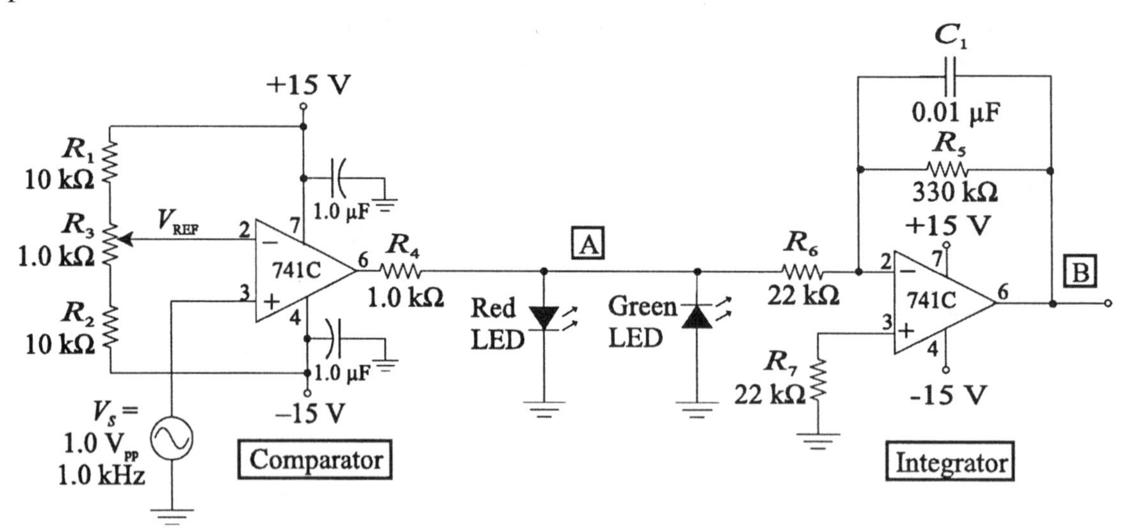

Figure 24-3

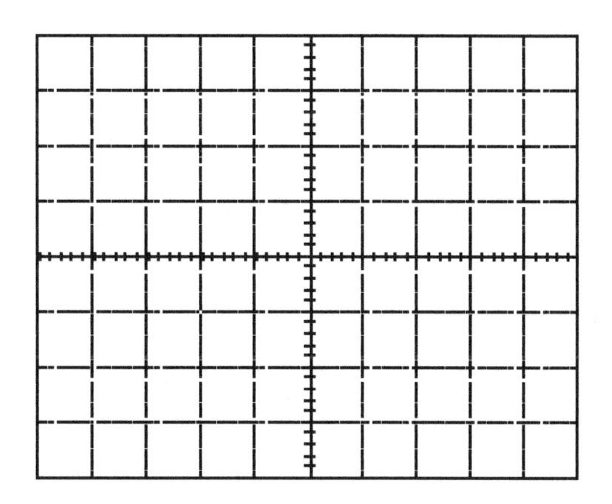

Plot 24-1

3. Vary R_3 while observing the output of the comparator and the integrator.

   Observations: ____

4. For each of the troubles listed in Table 24-2, see if you can predict the effect on the circuit. Then insert the trouble and check your prediction. At the end of this step, restore the circuit to normal operation.

Table 24-2

Trouble	Symptoms
No negative power supply	
Red LED open	
C_1 open	
R_5 open	

5. In this step, you will add a differentiating circuit to the previous circuit. The circuit is shown in Figure 24-4. Connect the input of the differentiator to the output of the integrator (point **B**). Observe the input and output waveforms of the differentiator. Sketch the observed waveforms on Plot 24-2. Show the voltages and time on your plot.

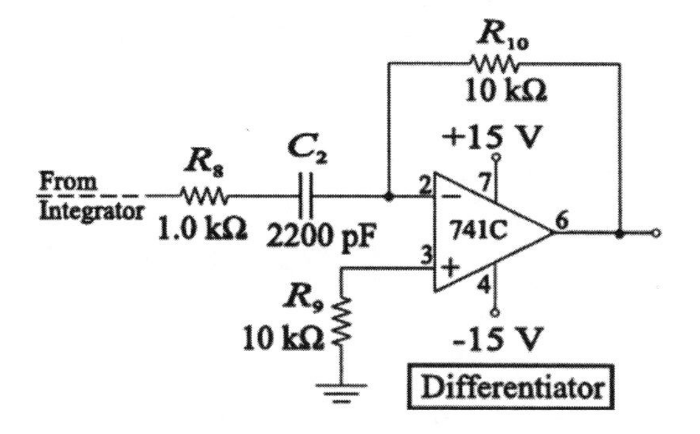

Figure 24-4

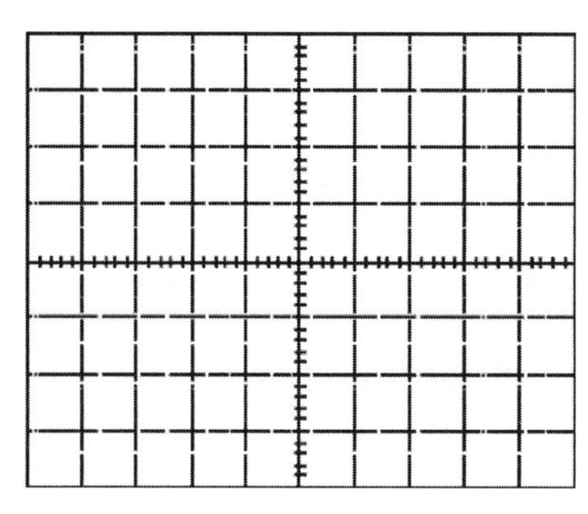

Plot 24-2

6. Remove the input from the differentiator and connect it to the output from the comparator (point **A**). Observe the new input and output waveforms of the differentiator. Sketch the observed waveforms on Plot 24-3. Show the voltages and time on your plot.

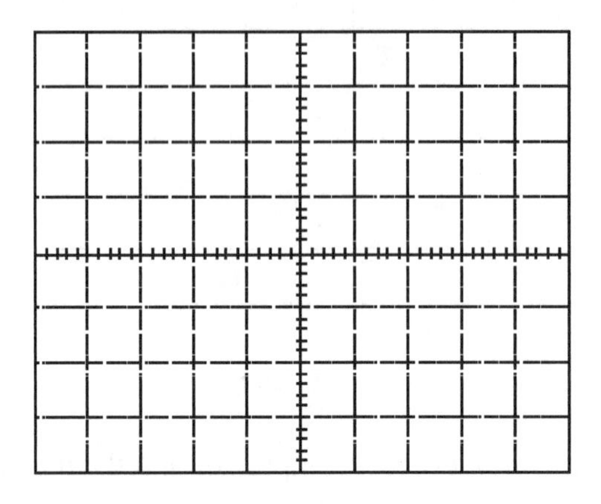

Plot 24-3

Conclusion:

Evaluation and Review Questions:

1. Compute the minimum and maximum V_{REF} for the comparator in Figure 24-2.

   $V_{REF(MIN)}$ = _______________ $V_{REF(MAX)}$ = _______________

2. The comparator output did not go near the power supply voltages. Explain why not.

3. (a) For the integrator circuit in Figure 24-3, what is the purpose of R_5?

   (b) What happened when it was removed?

4. What type of circuit will produce leading-edge and trailing-edge triggers from a square wave input?

5. What effect would you expect on the output of the integrator in Figure 24-3 if the frequency used had been 100 Hz instead of 1 kHz?

For Further Investigation:

In this experiment, you observed that an op-amp integrator produces the *inverse* of the true mathematical process of integration or summation. True integration (no inversion) occurs at the beginning of an *RC* charging circuit across the capacitor, but only if the charging voltage is much larger than the voltage on the capacitor. This situation can be made to occur with a Schmitt trigger oscillator, as long as the positive feedback is no more than about 10%. The circuit shown in Figure 24-5 will produce true integration of the square wave output of the Schmitt trigger on *C*. (Note that power supply voltages are not shown; the appropriate pins should be connected to ±15 V as in the experiment).

The Schmitt trigger was analyzed earlier (Experiment 22 – For Further Investigation), but a twist is added to this For Further Investigation by feeding the integrated output into a comparator similar to the one from this experiment. The output from the capacitor represents the integral of the square wave; it is used to control the duty cycle of a pulse waveform generated by the comparator. Investigate the circuit, measure the frequency and range of the duty cycle control. Observe the waveform on the capacitor and compare it to the output of the comparator. When does the output occur with respect to the triangle? The circuit can generate three waveforms simultaneously. What are they? Summarize your findings in a short report.

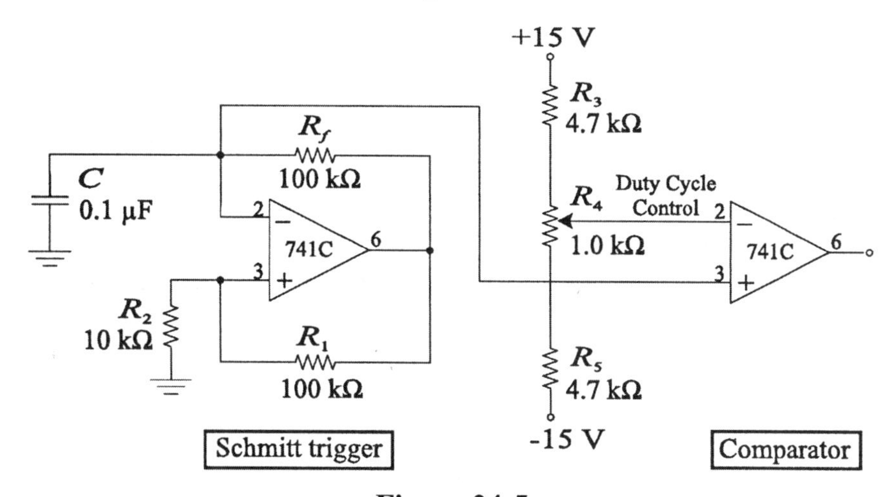

Figure 24-5

Name ______________
Date ______________
Class ______________

25 Low-Pass and High-Pass Active Filters

Reading:
Floyd and Buchla, *Analog Fundamentals: A Systems Approach*, Sections 9-1 through 9-4

Objectives:
After performing this experiment, you will be able to:
1. Specify the components required for a Butterworth low-pass or high-pass filter of a given order.
2. Build and test a Butterworth low-pass active filter for a specific frequency and order.
3. *For Further Investigation*: Change the low-pass filter into a high-pass design and test the response.

Summary of Theory:
A filter is a circuit that produces a prescribed frequency response. Passive filters are combination circuits containing only resistors, inductors, and capacitors (*RLC*). Active filters contain these elements and an operational amplifier (or transistor). The major advantage of active filters is that they can achieve frequency response characteristics that are nearly ideal and for reasonable cost for frequencies up to about 100 kHz. Above this, active filters are limited by bandwidth.

Active filters can be designed to optimize any of several characteristics. These include flatness of the response in the passband, steepness of the transition region, or minimum phase shift. The Butterworth form of filter has the flattest passband characteristic, but is not as steep as other filters and has poor phase characteristics. Since a flat passband is generally the most important characteristic, it will be used in this experiment.

The *order* of a filter, also called the number of *poles*, governs the steepness of the transition outside the frequencies of interest. In general, the higher the order, the steeper the response. The roll-off rate for active filters depends on the type of filter but is approximately –20 dB/decade for each pole. (A *decade* is a factor of ten in frequency). A four-pole filter, for example, has a roll-off of approximately –80 dB/decade. A quick way to determine the number of poles is to count the number of capacitors that are used in the frequency-determining part of the filter.

Figure 25-1 illustrates a two-pole active low-pass and a two-pole active high-pass filter. Each of these circuits is a *section*. To make a filter with more poles, simply cascade these sections, but change the gains of each section according to the values listed in Table 25-1. The cutoff frequency will be given by the equation:

$$f_c = \frac{1}{2\pi RC}$$

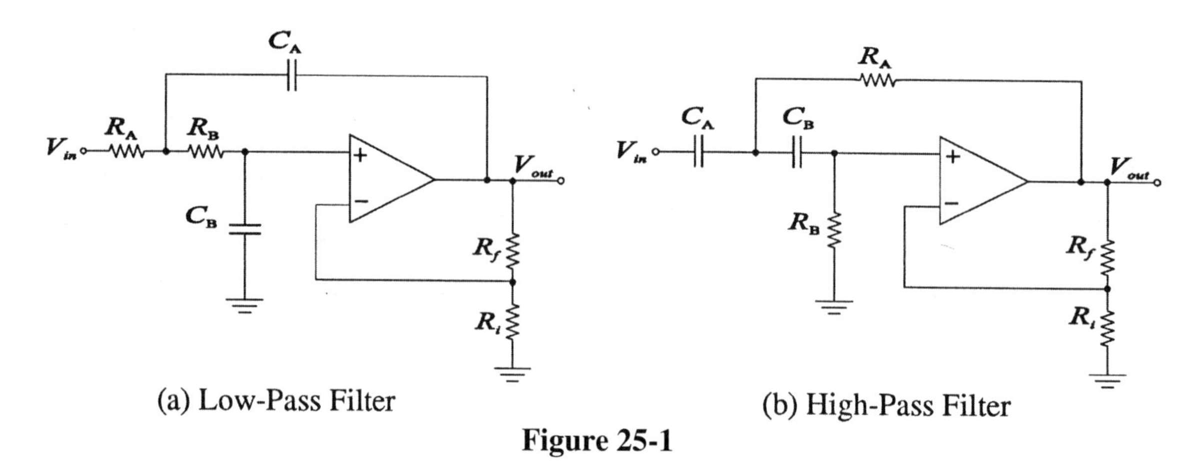

(a) Low-Pass Filter (b) High-Pass Filter

Figure 25-1

You can design your own Butterworth low-pass or high-pass active filter by using the following guidelines:

1. Determine the number of poles necessary based on the required roll-off rate. Choose an even number, as an odd number will require the same number of op-amps as the next highest even number. For example, if the required roll-off is –40 dB/decade, specify a two-pole filter.

2. Choose R and C values for the desired cutoff frequency (R_A, R_B, C_A and C_B). For best results, choose resistors between 1 kΩ and 100 kΩ. The values chosen should satisfy the cutoff frequency as given by the equation:

$$f_c = \frac{1}{2\pi RC}$$

3. Choose resistors R_f and R_i that give the gains for each section according to the values listed in Table 25-1. The gain is controlled only by R_f and R_i. Solving the closed-loop gain of a noninverting amplifier gives the equation for R_f in terms of R_i:

$$R_f = (A_v - 1)R_i$$

Table 25-1 Butterworth Low-Pass and High-Pass Filters

Poles	Gain Required		
	Section 1	Section 2	Section 3
2	1.586		
4	1.152	2.235	
6	1.068	1.586	2.483

<u>Example:</u> A low-pass Butterworth filter with a roll-off of approximately –80 dB/decade and a cutoff frequency of 2.0 kHz is required. Specify the components.

Step 1: Determine the number of poles required. Since the design requirement is for approximately –80 dB/decade, a four-pole (two-section) filter is required.

Step 2: Choose R and C. Try C as 0.01 μF and compute R. Computed R = 7.96 kΩ. Since the nearest standard value is 8.2 kΩ, choose C = 0.01 μF and R = 8.2 kΩ.

Step 3. Determine the gain required for each section and specify R_f and R_i. From Table 25-1, the gain of section 1 is required to be 1.152 and the gain of section 2 is required to be 2.235. Choose resistors that will give these gains for a noninverting amplifier. The choices are determined by again considering standard values and are shown on the completed schematic, Figure 25-2.

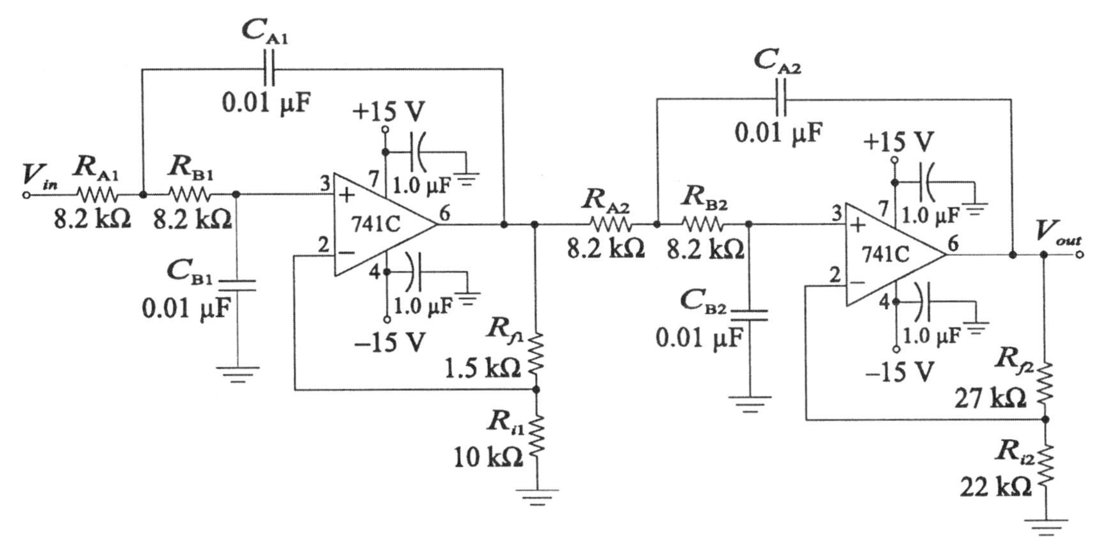

Figure 25-2

Materials Needed:

Resistors: one 1.5 kΩ, four 8.2 kΩ, two 10 kΩ, one 22 kΩ, one 27 kΩ
Capacitors: four 0.01 μF, four 1.0 μF
Two 741C op-amps

For Further Investigation:

One additional 741C op-amp and components to be specified by student

Procedure:

1. Measure and record the components listed in Table 25-2. If you cannot measure the capacitors, show the listed value.

Table 25-2

Component	Listed Value	Measured Values A1	B1	A2	B2
R_{A1} to R_{B2}	8.2 kΩ				
C_{A1} to C_{B2}	0.01 μF				
R_{i1}	10 kΩ				
R_{f1}	1.5 kΩ				
R_{i2}	22 kΩ				
R_{f2}	27 kΩ				

2. Construct the four-pole low-pass active filter illustrated in Figure 25-2. Install a 10 kΩ load resistor. Connect a sine wave generator to the input. Set it for a 500 Hz sine wave at 1.0 V rms. The voltage should be measured at the generator with the circuit connected. Set the voltage with a voltmeter and check both voltage and frequency with the oscilloscope. Measure V_{RL} at a frequency of 500 Hz, and record it in Table 25-3.

3. Change the frequency of the generator to 1000 Hz. Readjust the generator's amplitude to 1.0 V rms. Measure V_{RL}, entering the data in Table 25-3. Continue in this manner for each frequency listed in Table 25-3.

4. Graph the voltage across the load resistor (V_{RL}) as a function of frequency on Plot 25-1.

Table 25-3

Frequency	V_{RL}
500 Hz	
1000 Hz	
1500 Hz	
2000 Hz	
3000 Hz	
4000 Hz	
8000 Hz	

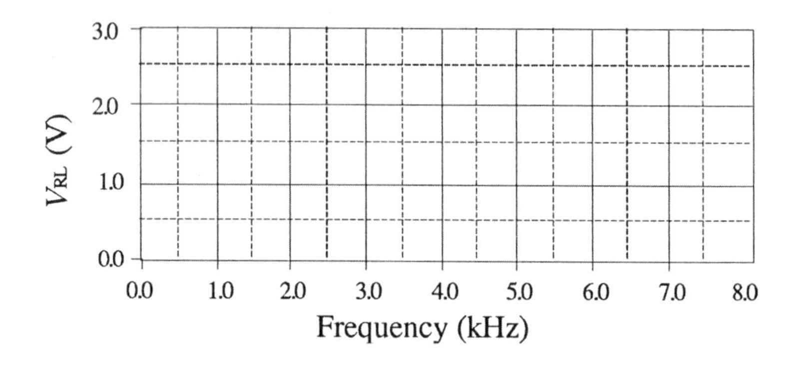

Plot 25-1

5. A Bode plot is a log-log plot of voltage versus frequency. It allows you to examine the data over a larger range than is possible with linear plots. Replot the data from the filter onto the log-log plot shown in Plot 25-2.

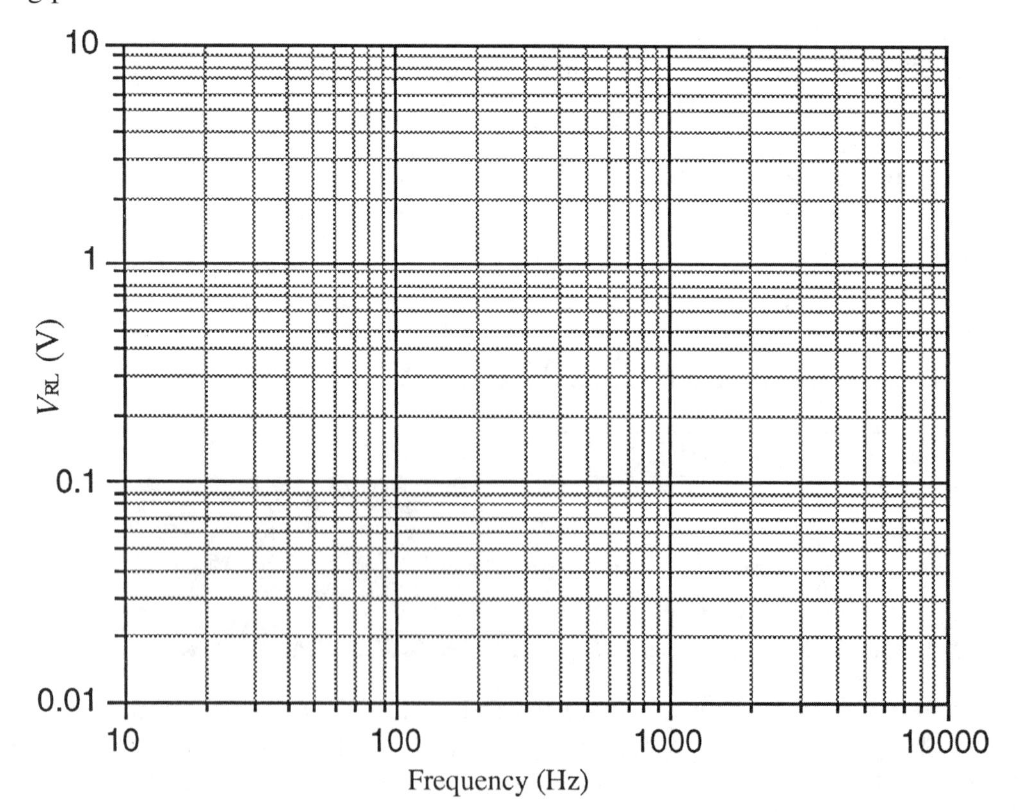

Plot 25-2

Conclusion:

Evaluation and Review Questions:

1. (a) From the frequency response curves, determine the cutoff frequency for the filter in this experiment.

   (b) Compute the average *R* and *C* for your active filter (Table 25-1). Use the average values of each to compute the cutoff frequency.

2. (a) What is the measured voltage gain of active filter in the passband?

   (b) What should it be?

3. Using the Bode plot, predict V_{out} at a frequency of 20 kHz.

4. The theoretical roll-off for your filter is –80 dB/decade. How does your actual filter compare to this theoretical roll-off rate?

5. (a) Using the measured values of R_{i1} and R_{f1}, compute the actual gain of the first section. Compare this to the required gain in Table 25-1.

   (b) Repeat for the second section using R_{i2} and R_{f2}.

For Further Investigation:
To convert from a Butterworth low-pass to a high-pass design requires only that the resistors and capacitors that make up the filter be exchanged (see Figure 25-1). Change the filter in the experiment to a high-pass filter and test the response. Submit a report summarizing your findings.

Name ______________
Date ______________
Class ______________

26 State-Variable Band-Pass Filter

Reading:
Floyd and Buchla, *Analog Fundamentals: A Systems Approach*, Section 9-5

Objectives:
After performing this experiment, you will be able to:

1. Construct a state-variable band-pass filter, measure and plot its frequency response.
2. Compute and measure the center frequency and Q of the band-pass filter.
3. Test the low-pass and high-pass outputs of the state-variable filter.
4. Describe how a swept frequency measurement is accomplished.
5. *For Further Investigation:* Set up a swept frequency circuit and display the frequency response on an oscilloscope.

Summary of Theory:
The previous experiment introduced active filters and showed how a VCVS filter could be designed as a low-pass or high-pass filter by simply setting the appropriate gain in a standard circuit. Although VCVS filters are popular and easy to construct, they normally cannot be used in high Q designs so they do not form outstanding band-pass filters. The main advantages of state-variable filters are that they can be designed for high Q (over 100) band-pass filters and they can be tuned to different frequencies or the bandwidth can be adjusted after the filter has been constructed. Because of the extra complexity, state-variable filters are usually used as band-pass filters, but they also have separate low-pass and high-pass outputs available from the same filter (band reject is also possible with an extra op-amp). In addition, state-variable filters are available in IC packages with internal capacitors that allow the user flexibility by choosing only external resistors to control gain, bandwidth, and Q.

The basic three op-amp state-variable filter uses two integrators connected in series. The band-pass output is taken from the first integrator, and the output of the second integrator is fed back to a summing amplifier as shown in Figure 26-1. Low-pass and high-pass outputs are available as shown. The equations for design of the filter come from network theory. For the simplified case shown, the center frequency is set by the RC time constant of the integrators. Assuming the same time constant for both integrators, the center frequency is found from:

$$f_0 = \frac{1}{2\pi R_4 C_1} = \frac{1}{2\pi R_7 C_2}$$

The Q is determined by the gain of the first integrator as follows:

$$Q = \frac{1}{3}\left(\frac{R_5}{R_6} + 1\right)$$

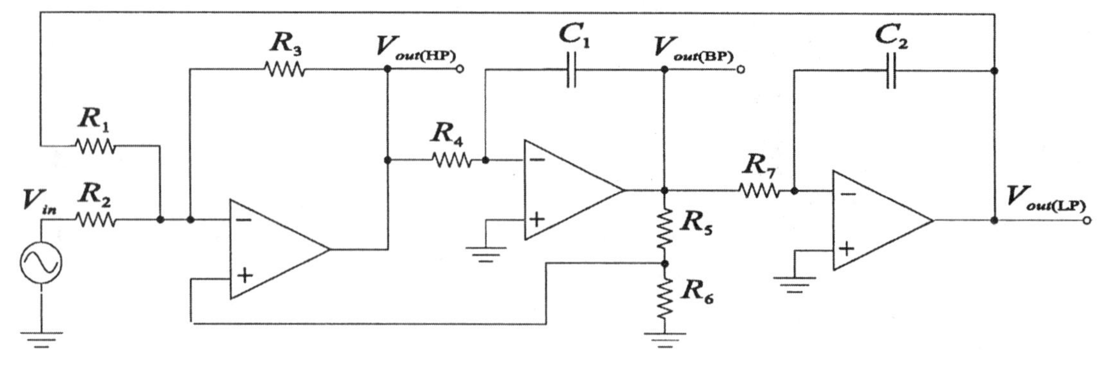

Figure 26-1

For certain applications, it is useful to convert the oscilloscope's time axis to a frequency axis, thus showing the frequency response on an oscilloscope. The easiest way to accomplish this is with a swept frequency generator; this is a generator that the frequency varies at a specific rate. The block diagram for a swept frequency measurement is shown in Figure 26-2. The generator's frequency changes at a rate determined by the sweep which is set to a very slow repetition rate. The sweep is brought out of the generator and used to synchronize the horizontal beam position on the scope. This causes the x-axis of the scope to represent frequency instead of time. This is particularly useful for viewing and adjusting a frequency sensitive circuit such as a tuned amplifier or a filter.

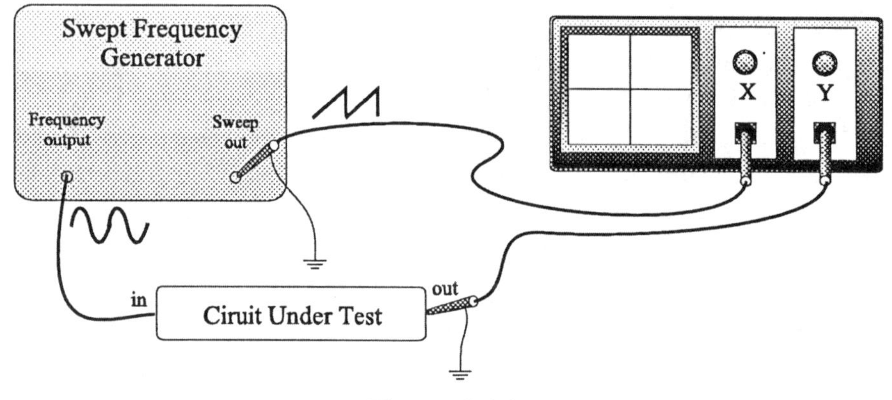

Figure 26-2

Materials Needed:

Resistors: three 1.0 kΩ, three 10 kΩ, one 100 kΩ
Capacitors: two 0.1 μF, two 1.0 μF
Three 741C op-amps

For Further Investigation:

Swept frequency generator for audio frequencies.

Procedure:

1. Measure and record the values of the resistors and capacitors listed in Table 26-1.

Table 26-1

Component	Listed Value	Measured Value
R_1	10 kΩ	
R_2	10 kΩ	
R_3	10 kΩ	
R_4	1.0 kΩ	
R_5	100 kΩ	
R_6	1.0 kΩ	
R_7	1.0 kΩ	
C_1	0.1 F	
C_2	0.1 F	

Table 26-2

Quantity	Computed	Measured
Center frequency, f_0 =		
$V_{pp(center)}$ =		
Upper cutoff, f_{cu} =		
Lower cutoff, f_{cl} =		
Bandwidth, BW =		
Q =		

2. For the state-variable filter shown in Figure 26-3, compute the center frequency, f_0, and the Q of the circuit from the equations given in the Summary of Theory. Compute the bandwidth, BW, by dividing f_0 by the Q. Enter these computed values in Table 26-2.

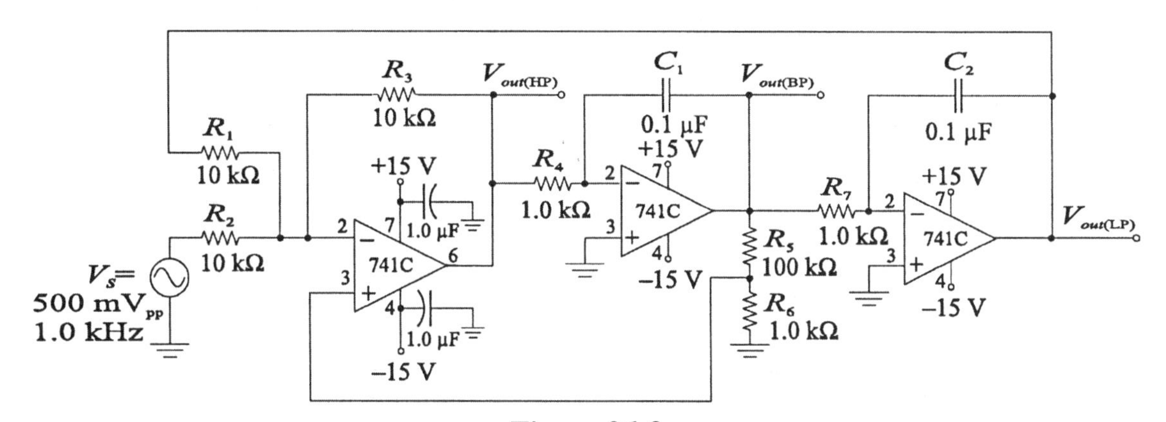

Figure 26-3

3. Construct the circuit shown in Figure 26-3. You will need the 1.0 μF bypass capacitors on the power supplies for only one op-amp, as shown. Set the function generator for a 500 mV$_{pp}$ sine wave at 1.0 kHz. Test the response from the band-pass output (center op-amp) by slowly increasing the frequency of the function generator as you observe the response. You should observe a dramatic peak in the output at the center frequency.

4. Observe the output with the oscilloscope and tune the center frequency, f_0 (maximum output). Measure this frequency and the peak-to-peak output voltage, $V_{pp(center)}$. Then carefully vary the frequency above and below the center until the output drops to 70.7% of the maximum voltage. These are the upper and lower cutoff frequencies, f_{cu} and f_{cl}, for the filter. The measured bandwidth is the difference between f_{cl} and f_{cu}. Record the measured values in Table 26-2. If available, you can obtain better accuracy if you use a frequency counter for the frequency measurements.

5. To obtain a better idea of the frequency response of your filter, measure and record the peak-to-peak output voltage as a function of frequency for the values listed in Table 26-3. Plot the response in Plot 26-1. Include the voltage for the center frequency on your plot. Because of the large dynamic range of the data, the plot is logarithmic.

Table 26-3

Frequency	Output voltage, V_{pp}
100 Hz	
200 Hz	
500 Hz	
1.0 kHz	
1.5 kHz	
2.0 kHz	

Table 26-3 (continued)

Frequency	Output voltage, V_{pp}
2.5 kHz	
3.0 kHz	
4.0 kHz	
5.0 kHz	
10 kHz	
20 kHz	

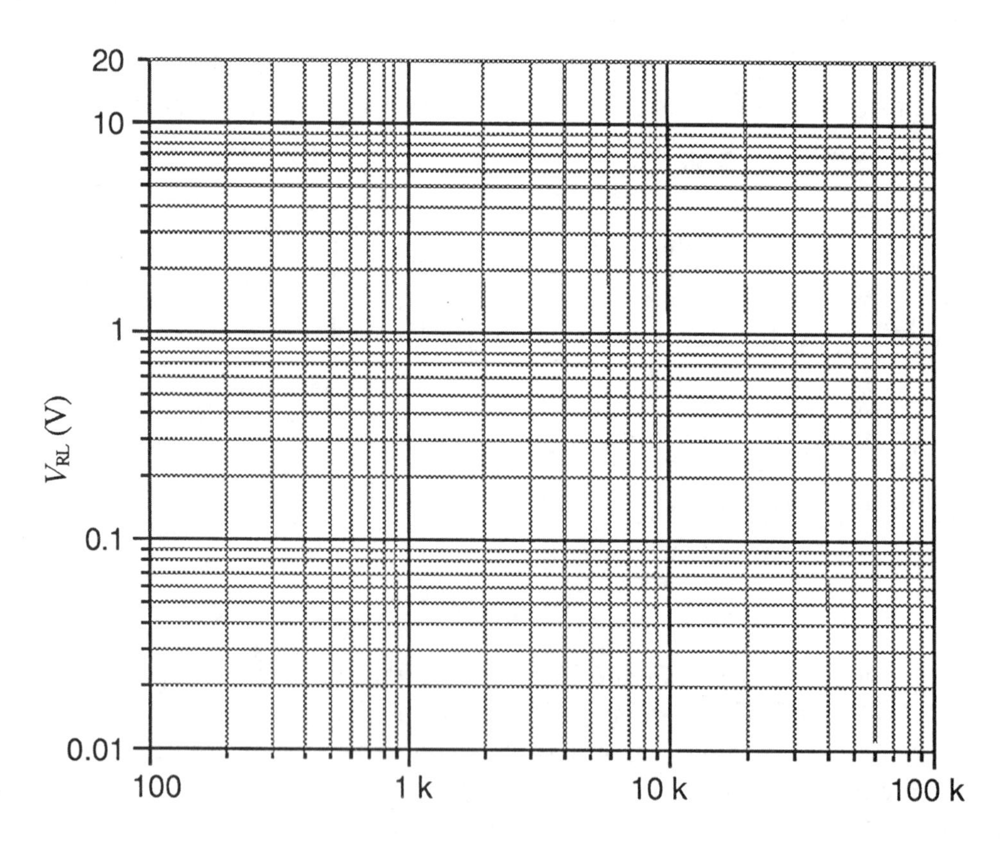

Plot 26-1

6. In addition to the band-pass output, the state-variable filter has low-pass and high-pass outputs as shown on Figure 26-3. Test these outputs in the circuit you constructed. What is the gain in the passband? Notice the gain change near the cutoff frequency. Can you think of a way of eliminating the peaking that occurs?

   Observations:____

   ____

Conclusion:

Evaluation and Review Questions:

1. The input signal for the filter was set for only 500 mV_{pp}. Under what circumstances could a larger signal be used without driving the output into saturation?

2. (a) Give the principal advantages of the state-variable filter in this experiment over the VCVS filter in the previous experiment.

   (b) One disadvantage of the state-variable filter in this experiment used as a high-pass or low-pass filter is the gain change near the cutoff frequency that you observed in step 6. What change would you suggest to eliminate this problem? Does this affect the band-pass filter?

3. (a) What change would you make to the circuit in this experiment to raise the center frequency by a factor of two?

   (b) What change would you make to lower the Q of the circuit?

4. (a) Assume the inputs of a summing amplifier are connected to the low-pass and high-pass outputs of the state-variable filter in this experiment. What type of filter does this form?

   (b) Name an application for this type of filter.

5. Refer to the block diagram of the swept frequency measurement in Figure 26-2. What are *two* functions of the sawtooth voltage from the swept frequency generator?

For Further Investigation:

As described in the Summary of Theory, a swept frequency measurement enables you to view the frequency response of the filter directly on the display of the oscilloscope. If you have a swept frequency generator, set it up as shown in the block diagram in Figure 26-2. Use a slow repetition rate (about 1 Hz) and adjust the output of the generator for a 500 mV_{pp} signal. Use dc coupling on your oscilloscope. You should be able to adjust the scope to show the frequency response of your filter. Write up a summary of your observations.

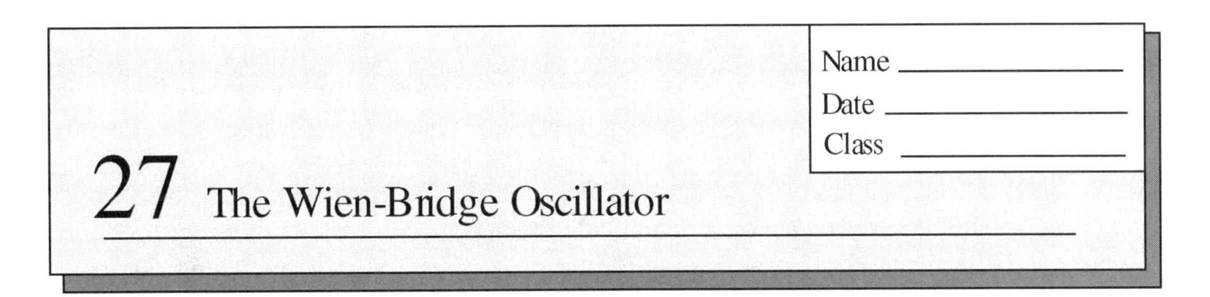

Reading:
Floyd and Buchla, *Analog Fundamentals: A Systems Approach*, Sections 10-1 through 10-3

Objectives:
After performing this experiment, you will be able to:

1. Explain the requirements for a Wien bridge to oscillate, predict the feedback voltages and phases, and compute the frequency of oscillation.
2. Construct and test a FET stabilized Wien-bridge oscillator.

Summary of Theory:
The Wien bridge is a bridge-type circuit that is widely used as a sinusoidal oscillator for frequencies below about 1 MHz. Oscillation occurs when a portion of the output is returned to the input in the proper amplitude and phase to reinforce the input signal. This type of feedback is called *regenerative* or *positive* feedback. All regenerative oscillators require amplification to overcome the loss in the feedback network. For the Wien bridge, the feedback network returns 1/3 of the output signal to the noninverting input. Therefore, the amplifier must provide a gain of 3 to overcome this attenuation and prevent oscillations from dying out.

The basic Wien-bridge circuit is shown in Figure 27-1(a). The gain is controlled by the resistors connected to the inverting input of the op-amp (R_f and R_i). The frequency of oscillation is determined by the lead-lag network connected to the noninverting input of the op-amp. The oscillation frequency is found from the equation:

$$f_r = \frac{1}{2\pi RC}$$

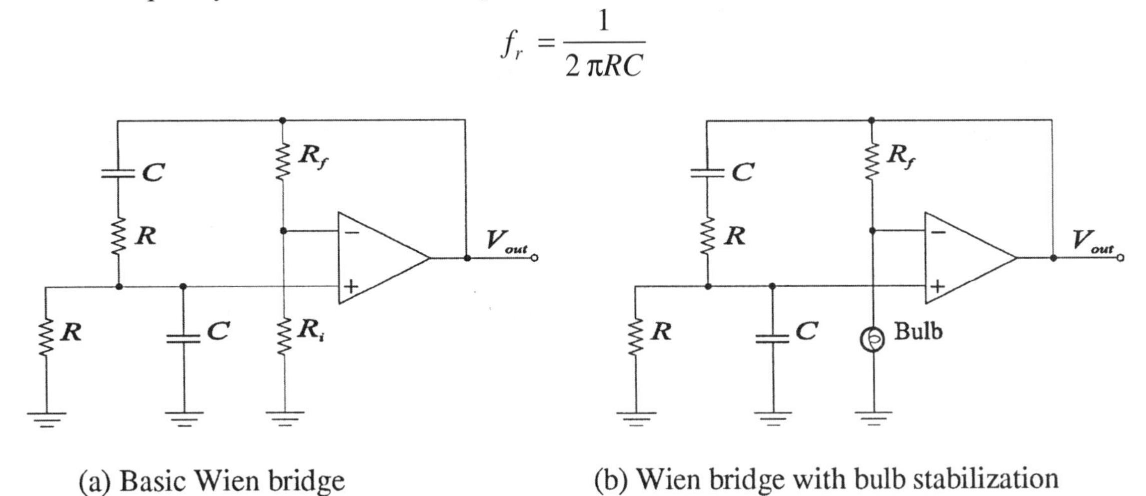

(a) Basic Wien bridge (b) Wien bridge with bulb stabilization

Figure 27-1

The gain must be at least 3 to maintain oscillations but too much gain causes the output to saturate. Too little gain causes oscillations to cease. Various circuits have been designed to stabilize loop gain at the required 3. The basic requirement is to provide *automatic gain control*, or AGC for short. One common technique is to use a light bulb that has a positive temperature coefficient for AGC as illustrated in Figure 27-1(b). As the bulb's filament warms, its resistance increases and reduces the gain. Other more sophisticated techniques use the ohmic region of a FET as a variable resistor to control gain. You will investigate a FET stabilized Wien bridge in this experiment.

Materials Needed:
Resistors: one 1.0 kΩ, three 10 kΩ
Capacitors: two 0.01 μF, three 1.0 μF
Two 1N914 signal diodes (or equivalent)
One 741C op-amp
One 2N5458 *n*-channel JFET transistor (or equivalent)
One 10 kΩ potentiometer

For Further Investigation:
One type #1869 or type #327 bulb

Procedure:

1. Measure R_1, R_2, C_1, and C_2 for this experiment. These components determine the frequency of the Wien bridge. Record the measured values in Table 27-1. If you cannot measure the capacitors, record the listed value.

2. Construct the basic Wien bridge illustrated in Figure 27-2. Adjust R_f so that the circuit just oscillates. You will see that it is nearly impossible to obtain a clean sine wave as the control is too sensitive. With the bridge oscillating, try spraying some freeze spray on the components and observe the result.

   Observations:____

Table 27-1

Component	Listed Value	Measured Value
R_1	10 kΩ	
R_2	10 kΩ	
C_1	0.01 F	
C_2	0.01 F	

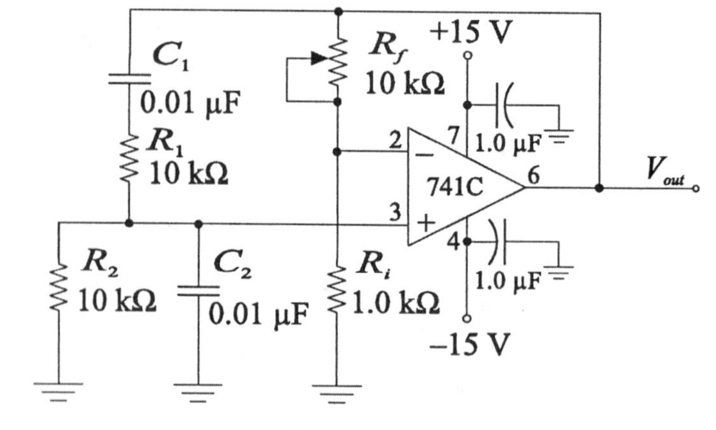

Figure 27-2

3. The bridge in step 2 has the problem of unstable gain and requires some form of automatic gain control to work properly. Field-effect transistors are frequently used for AGC circuits because they can be used as voltage-controlled resistors for small applied voltages (see text discussion on page 206). The circuit illustrated in Figure 27-3 is a FET-stabilized Wien bridge. Compute the expected frequency of oscillation from the equation:

$$f_r = \frac{1}{2\pi RC}$$

Use the average measured value of the resistance and capacitance listed in Table 27-1 to calculate f_r. Record the computed f_r in Table 27-2.

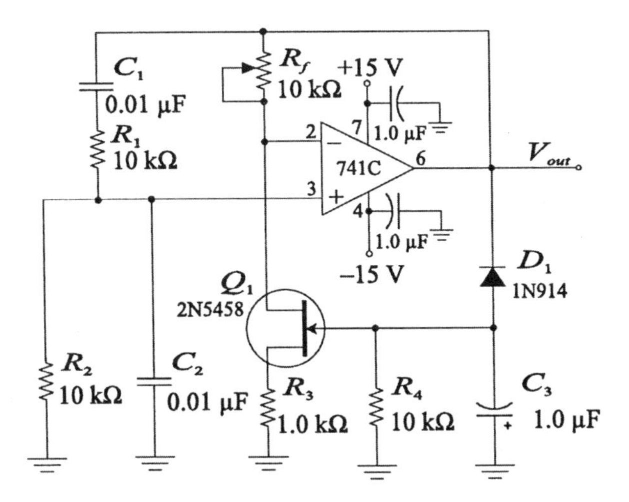

Figure 27-3

Table 27-2

f_r Computed	f_r Measured (pin 6)

4. Construct the FET-stabilized Wien bridge shown in Figure 27-3. The diode causes negative peaks to charge C_3 and bias the FET. C_3 has a relatively long time constant discharge path (through R_4) so the bias does not change rapidly. Note the polarity of C_3. Adjust R_f for a good sine wave output. Measure the frequency and record it in Table 27-2.

5. Measure the peak-to-peak output voltage, $V_{out(pp)}$. Then measure the peak-to-peak positive and negative feedback voltages, $V_{(+)(pp)}$ and $V_{(-)(pp)}$ and the dc voltage on the gate of the FET (V_G). Use two channels and observe the phase relationship of the waveforms. Record voltages in Table 27-3.

Table 27-3

Measured Voltages			
$V_{out(pp)}$ (pin 6)	$V_{(+)(pp)}$ (pin 3)	$V_{(-)(pp)}$ (pin 2)	V_G

What is the phase shift from the output voltage to the positive feedback voltage? _______

6. Try freeze spray on the various components while observing the output.

Observations:____

7. Add a second diode in series with the first one between the output and the gate of the FET (See Figure 27-4). You may need to readjust R_f for a good sine wave. Measure the voltages as before and record in Table 27-4.

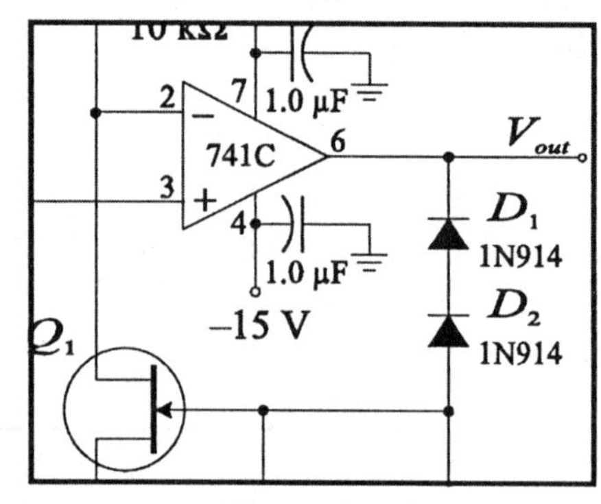

Figure 27-4

Table 27-3

Measured Voltages			
$V_{out(pp)}$ (pin 6)	$V_{(+)(pp)}$ (pin 3)	$V_{(-)(pp)}$ (pin 2)	V_G

Conclusion:

Evaluation and Review Questions:

1. In step 5, you measured the positive feedback voltage.
   (a) What fraction of the output voltage did you find?

   (b) Is this what you expect from theory?

2. Explain why adding a second diode in series with the first caused the output voltage to increase.

3. For the circuit in Figure 27-3, why is the positive side of C_3 shown on ground?

4. What frequency would the Wien bridge of Figure 27-3 oscillate if R_1 and R_2 were doubled?

5. How could you make a Wien bridge tune to different frequencies?

For Further Investigation:

Investigate the light bulb stabilized Wien bridge shown in Figure 27-1(b). A good bulb to try is a type #1869 or type #327. Other bulbs will work, but low-resistance filaments are not good. You can use the same components as in Figure 27-2 except replace R_3 with the bulb. Summarize your results in a short lab report.

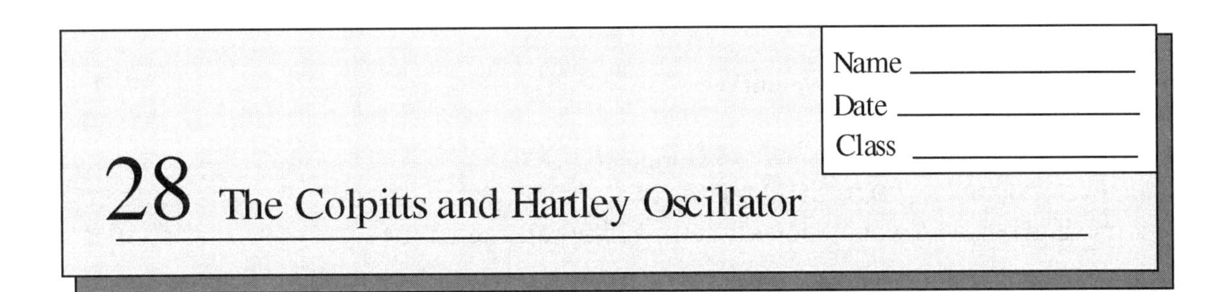

Reading:
Floyd and Buchla, *Analog Fundamentals: A Systems Approach*, Sections 10-1 through 10-4

Objectives:
After performing this experiment, you will be able to:

1. Connect a class-A amplifier; calculate and measure the dc and ac parameters.
2. Modify the amplifier with a feedback circuit that forms two versions of *LC* oscillators – the Colpitts and the Hartley.
3. Compare the computed and measured performance of the oscillators.

Summary of Theory:
In electronic systems, there are many requirements for one or more circuits that generate a continuous waveform. Several of these circuits are discussed in the text. A free-running oscillator is basically an amplifier that generates a continuous alternating voltage by feeding a portion of the output signal back to the input. The circuit in this experiment is based on the principles of a feedback oscillator, but introduces two other ways of obtaining the feedback, not discussed in the text.

Sinusoidal oscillators are classified by the networks used to provide feedback. To sustain oscillations, the amplifier must have sufficient gain to overcome the losses in the feedback network. In addition, the feedback must be of the proper phase to ensure that the signal is reinforced at the output – in other words, there must be *positive* feedback. One example is the phase shift oscillator that uses an *RC* network to produce the proper feedback (see Section 10-3 of the text). Feedback networks can be classified as *RC, LC,* or by a *crystal*, a special piezoelectric resonant network. In the previous experiment, you generated a sine wave with a Wien bridge that uses an *RC* feedback network. The Wien bridge is an excellent choice for frequencies below 1 MHz. For higher frequencies, *LC* circuits and crystal oscillators are generally preferred.

LC circuits have a parallel resonant circuit, commonly referred to as the *tank* circuit, that determines the frequency of oscillation (see Section 10-4 of the text). A portion of the output is returned to the input causing the amplifier to conduct only during a very small part of the total period. This means that the amplifier is actually running in class C mode. In applications where frequency stability is important, crystal oscillators have the advantage. In this experiment, you will test two *LC* oscillators and, in the For Further Investigation section, you will test a crystal oscillator.

Materials Needed:
One 2N3904 *npn* transistor (or equivalent)
One 100 Ω potentiometer
Resistors: one 1.0 kΩ, one 2.7 kΩ, one 3.3 kΩ, one 10 kΩ
Capacitors: two 1000 pF, one 0.01 μF, three 0.1 μF
Inductors: one 2 μH (can be wound quickly from #22 wire), one 25 μH

For Further Investigation:
One 1.0 MHz crystal
One 2N5458 *n*-channel JFET
One 1.0 MΩ resistor

Procedure:

1. Measure and record the value of the resistors listed in Table 28-1.
2. Observe the class A amplifier shown in Figure 28-1. Using your measured resistor values, compute the dc parameters for the amplifier listed in Table 28-2. R_{E1} is a 100 Ω potentiometer that you should set to 50 Ω. Then construct the circuit and verify that your computed dc parameters are as expected. Record the measured values in Table 28-2.

Table 28-1

Resistor	Listed Value	Measured Value
R_1	10 kΩ	
R_2	3.3 kΩ	
R_{E1}	50 Ω *	
R_{E2}	1.0 kΩ	
R_C	2.7 kΩ	

* set potentiometer for 50 Ω

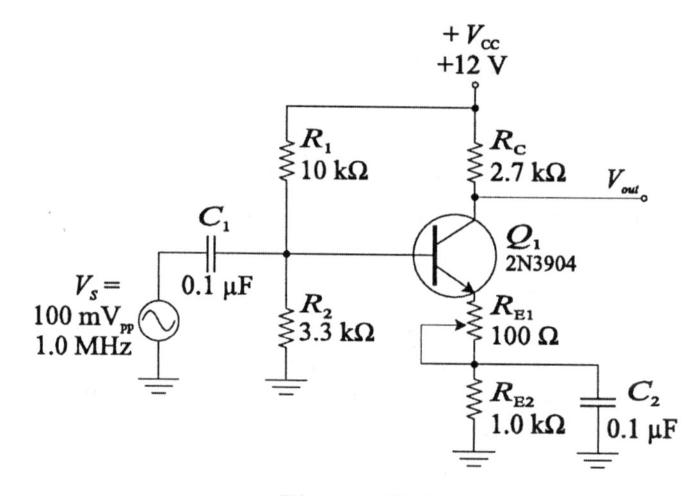

Figure 28-1

3. Compute the ac parameters listed in Table 28-3. After you find r_e', the gain is found by dividing the collector resistance by the sum of the unbypassed emitter resistance and r_e'. (Assume that the potentiometer remains set to 50 Ω.) The ac voltage at the collector is found by multiplying the gain by the ac base voltage. Set the function generator for a 100 mV$_{pp}$ signal at 1.0 MHz and measure the peak-to-peak collector voltage. Your computed and measured values should agree within normal experimental uncertainty.

Table 28-2

DC Parameter	Computed Value	Measured Value
V_B		
V_E		
I_E		
V_C		

Table 28-3

AC Parameter	Computed Value	Measured Value
V_b		
r_e'		
A_v		
V_c		

4. Remove the signal generator and add the feedback network for a Colpitts oscillator as shown in Figure 28-2. Adjust R_{E1} for the best sine wave. Compute the frequency of the Colpitts oscillator and record the computed frequency in Table 28-4. Then, measure the frequency and the peak-to-peak voltage at the output and record them in Table 28-4.

Table 28-4

Colpitts Oscillator	Computed Value	Measured Value
frequency		
amplitude		

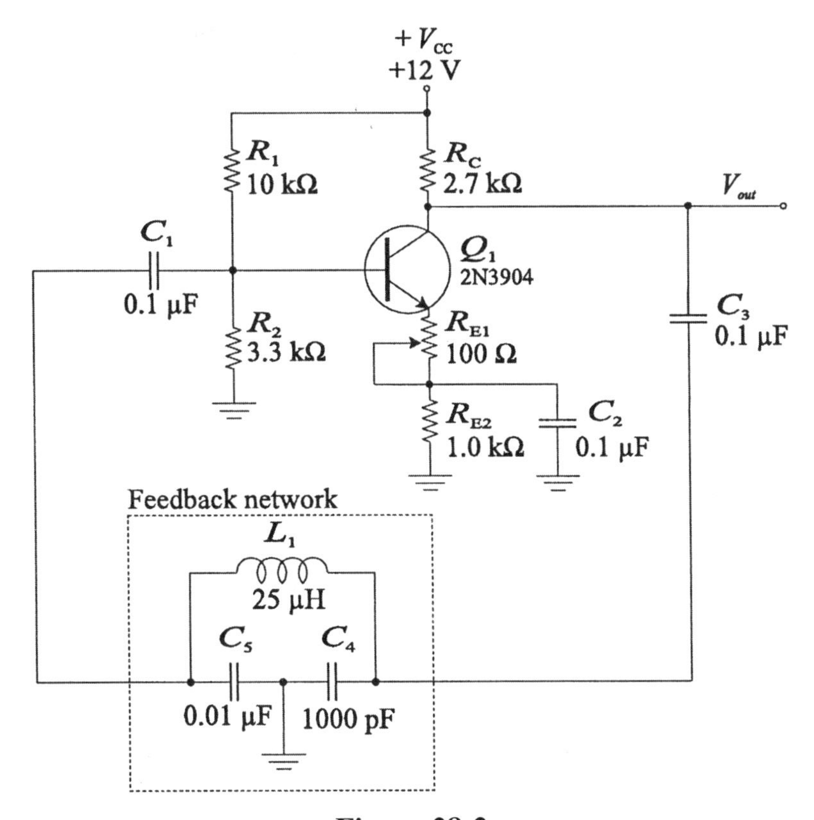

Figure 28-2

5. Observe what happens to the frequency and amplitude of the output signal when another 1000 pF capacitor is placed in parallel with C_4.

____

____

6. Observe the effect of freeze spray on the stability of the oscillator.

____

____

7. Replace the feedback network with the one shown in Figure 28-3. (L_2 can be wound by wrapping about 40 turns of #22 wire on a pencil). Adjust R_{E1} for a good sine wave output. This configuration is that of a Hartley oscillator. Compute the frequency of the Hartley oscillator, and record the computed frequency in Table 28-5. Then, measure the frequency and the peak-to-peak voltage at the output and record them in Table 28-5.

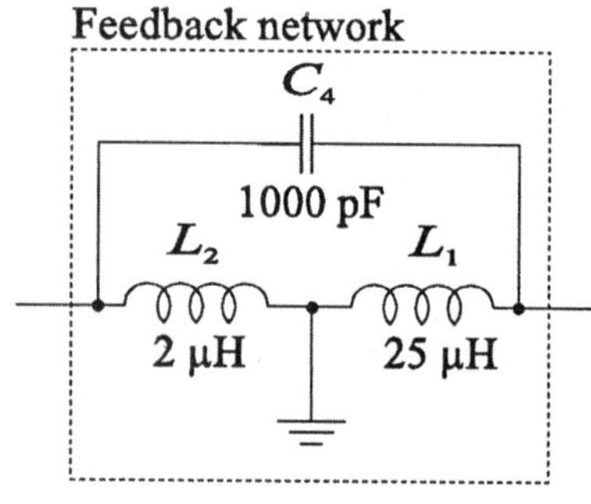

Figure 28-3

Table 28-5

Hartley Oscillator	Computed Value	Measured Value
frequency		
amplitude		

Conclusion:

Evaluation and Review Questions:

1. In step 5, you observed a change in the amplitude of the output signal when a capacitor was placed in parallel with C_4. Since the gain of the class A amplifier remained the same, what conclusion can you draw about the effect of the change on the amount of feedback?

2. What are the two conditions required for oscillation to occur in an *LC* oscillator?

3. Give a reason that an oscillator might drift from its normal frequency.

4. Summarize the difference between a Colpitts and Hartley oscillator.

5. For the circuit in Figure 28-2, predict the outcome in each case.

   (a) R_{E1} is shorted:

   (b) C_3 and C_4 are reversed:

   (c) C_2 is open:

   (d) The power supply voltage is +6 V:

For Further Investigation:

When it is necessary to have high stability in an oscillator, a crystal oscillator is superior. For high-frequency crystal oscillators, FETs have advantages over bipolar transistors because of their high input impedance. This allows the tank circuit to be unloaded, resulting in a high Q. The circuit shown in Figure 28-4 has the advantage of being simple, yet very stable. Construct the circuit and observe the waveform at the drain. Compare the frequency with that stamped on the crystal case (to do this requires a frequency counter). Test the effect of freeze spray on the frequency and amplitude. Summarize your results.

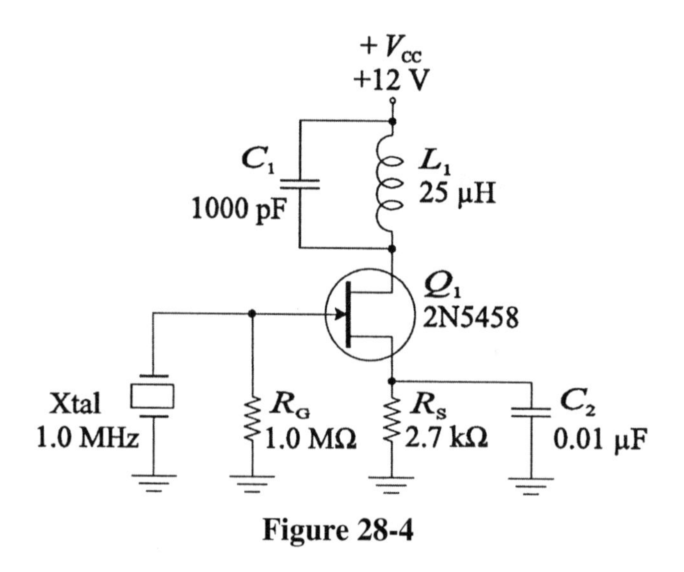

Figure 28-4

29 The 555 Timer

Reading:
Floyd and Buchla, *Analog Fundamentals: A Systems Approach*, Sections 10-5 through 10-7

Objectives:
After performing this experiment, you will be able to:

1. Calculate and measure the frequency and duty cycle for an astable 555 timer.
2. Test an astable circuit that uses a light sensor.
3. Calculate and measure the pulse width of a 555 timer configured as a one shot.

Summary of Theory:
The 555 timer was the first integrated circuit timer and is still one of the most popular because of its low price, wide range of times available (from microseconds to hours), and its versatility. It can be configured as either an astable (continuous pulses) or as a one-shot depending on the external circuit. In the astable configuration, it is a relaxation oscillator, using an *RC* timing circuit to control the frequency.

Internally, the 555 timer contains a voltage divider consisting of three series 5 kΩ resistors, two comparators, a control flip-flop, a discharge transistor, and an inverting output buffer, as shown in Figure 29-1. In normal operation (control input not used), the voltage divider sets a reference voltage on comparator *A* of 2/3 of V_{CC}. If the threshold input exceeds 2/3 of V_{CC}, comparator *A* output is high and sets the control flip-flop output high. On comparator *B*, the divider sets a reference voltage of 1/3 V_{CC}. If the trigger input goes below 1/3 of V_{CC}, the output of comparator *B* resets the control flip-flop output low.

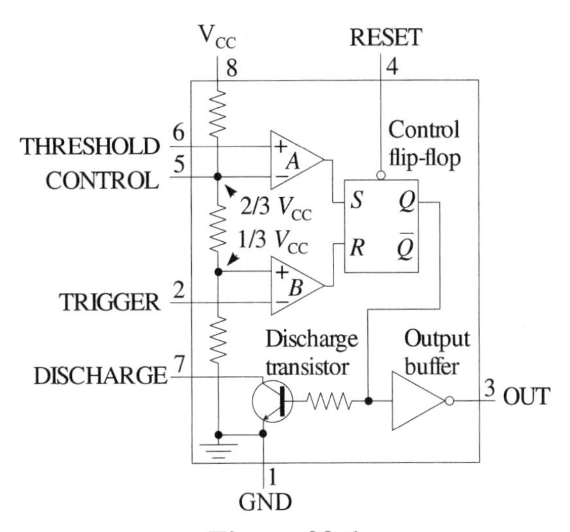

Figure 29-1

Other inputs include the *discharge*, *reset*, and *control* inputs. When the flip-flop is set, the Q output is high, causing the discharge transistor to be forward-biased, acting as a closed switch to ground through the discharge input. In the astable mode of operation, this provides a path through a resistor for discharging the timing capacitor. The reset input is normally connected to a logic high, but can be used to place the 555 output to a low state. The control input enables the user to change the reference voltage on the voltage divider, but for most applications, the control input is left open or connected through a small capacitor to ground. In this experiment, it will not be used.

An example of a basic astable circuit is shown in Figure 29-2(a). In this mode of operation, the equations for frequency and duty cycle are given in the text (Equations 10-12 and 10-13) and repeated here for reference.

$$f = \frac{1.44}{(R_1 + 2R_2)C_{\text{ext}}} \qquad \text{Duty cycle} = \left(\frac{R_1 + R_2}{R_1 + 2R_2}\right)100\%$$

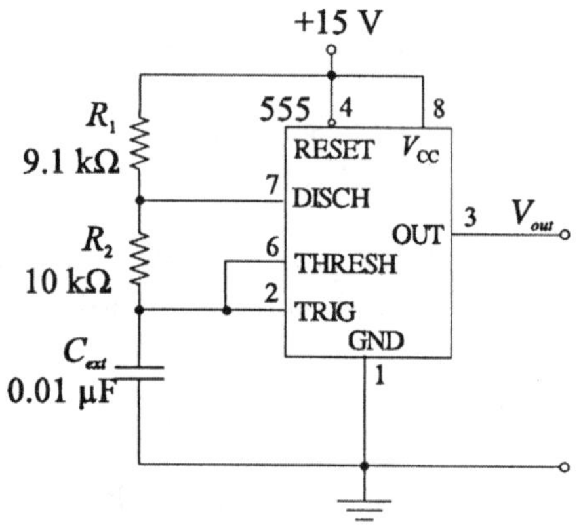

(a) Basic astable operation

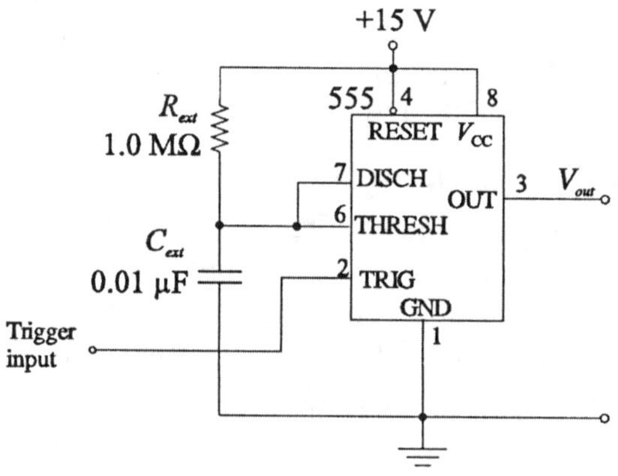

(b) Basic monostable operation

Figure 29-2

A basic monostable circuit is shown in Figure 29-2(b). Mechanical switches tend to produce multiple pulses when closed, a problem for many digital circuits. One application of this circuit is to produce a single short pulse for "debouncing" a switch as you will see in step 7. The switch is capacitively coupled with a pull-up resistor on each side to keep the line high when the switch is open. In the monostable mode, the equation for pulse width is as given in the text (Equation 10-15) and repeated here for reference:

$$t_w = 1.1\,R_{\text{ext}}\,C_{\text{ext}}$$

Materials Needed:

Resistors: two 1.0 kΩ, one 9.1 kΩ, one 10 kΩ, one 22 kΩ, one 1.0 MΩ
Capacitors: one 0.01 μF, one 0.1 μF
One 555 timer IC
One 1N914 diode (or equivalent)
One CdS photocell – Electronix Express 08GL7516 or equivalent

For Further Investigation:

One 100 kΩ potentiometer

Procedure:

1. Measure and record the values of the timing resistors and capacitor listed in Table 29-1.

Table 29-1

Resistor	Listed Value	Measured Value
R_1	9.1 kΩ	
R_2	10 kΩ	
C_{ext}	0.01 μF	

Table 29-2

Quantity	Computed Value	Measured Value
frequency		
duty cycle		

2. Compute the frequency and duty cycle of the astable circuit shown in Figure 29-2(a). Enter the computed values in Table 29-2.

3. Connect the circuit shown in Figure 29-2(a). Measure and record the frequency and duty cycle listed in Table 29-2.

4. With the circuit operating, observe the waveform across the capacitor, C_{ext}. Couple the oscilloscope with dc coupling and note the ground level for the signal. Sketch the waveform on Plot 29-1. Label the voltage and note the ground position on your sketch.

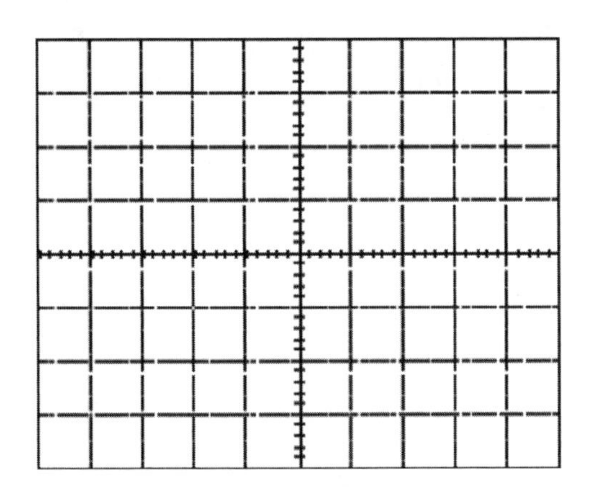

Plot 29-1

Figure 29-3

5. Connect a diode across R_2 as shown in Figure 29-3. The diode represents a path around R_2 for charging the capacitor but has no effect on the discharge path. Because of the diode's forward drop and temperature dependence, this circuit works best with the +15 V supply shown. Observe the effect on the frequency and the duty cycle. Try changing the supply voltage and observe what happens to the output.

Observations: ____

____

6. An interesting variation in this circuit is made by replacing R_1 with a CdS photocell. The CdS photocell changes resistance as light strikes it. For many transducers, converting the output to a variable frequency is a useful first step in processing the data.

   Replace R_1 with the photocell and observe the variation in the output frequency as you cover it with your hand.

   Observations: ____

7. Figure 29-4 shows a one-shot circuit to debounce a pushbutton switch. The switch is shown for reference only; you will simulate the switch with a pulse generator. The purpose of the input circuit (R_1, R_2, R_3 and C_1) is to assure the trigger conditions are met even if the trigger is smaller. Compute the pulse width and enter it in Table 29-3.

   Connect the circuit; then measure the output pulse width. To measure this relatively long pulse, replace the switch with a pulse generator set for a 0 to 5 V pulse at 5 Hz and obtain a stable display on an oscilloscope. It is difficult to obtain a stable display on an oscilloscope with a frequency this low. Try triggering the scope using *normal* triggering, and carefully adjust the trigger level for a stable display. Measure the pulse width and record it in Table 29-3.

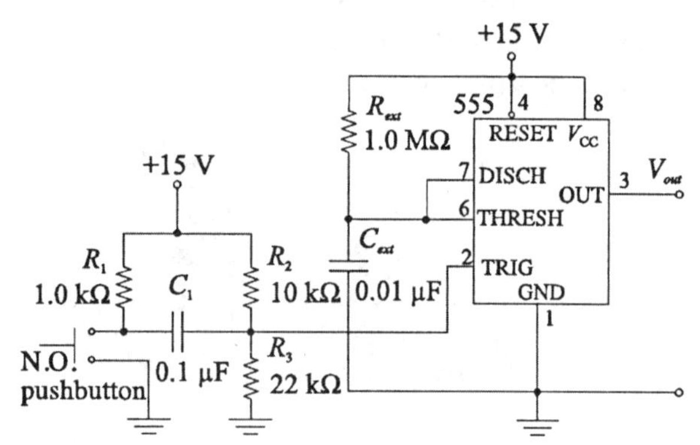

Figure 29-4

Table 29-3

Quantity	Computed Value	Measured Value
pulse width		

8. When you have achieved a stable trace for the circuit in step 7, try connecting the second channel of your oscilloscope across the capacitor. Sketch the waveforms on Plot 29-2.

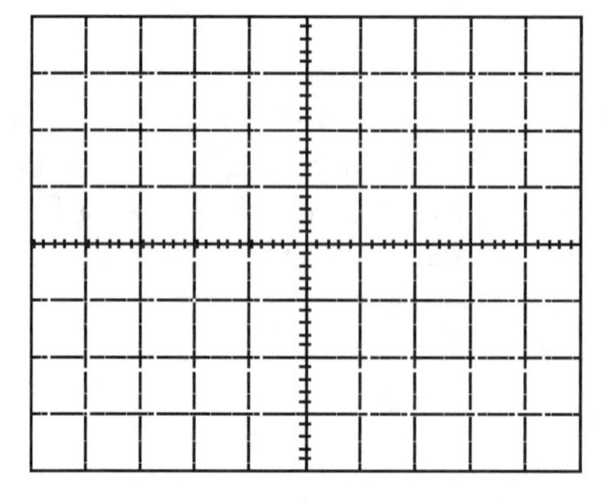

Plot 29-2

Conclusion:

Evaluation and Review Questions:

1. In the basic astable circuit drawn in Figure 29-2(a), it is not possible to obtain a 50% duty cycle. Explain why not.

2. (a) In step 4, you observed the voltage across the capacitor for the basic astable circuit. How would the amplitude of this voltage change if the power supply voltage were reduced to +12 V?

   (b) Would this have an effect on the frequency from the 555 timer? Why or why not?

3. What change to the astable circuit would you suggest to produce output pulses that were on for 1 s and off for 1 s? Draw the circuit and show how you could light an LED with the high output and another LED with the low output.

4. Look up the manufacturer's specification sheet for the 555 timer[1].
   (a) What is the maximum current the output can source or sink?

   (b) What happens to the output voltage when the timer supplies higher current?

5. Why was a frequency of 5 Hz suggested in step 7 for looking at the debounce circuit?

---

[1] The specification sheet is available from Texas Instruments at http://www.ti.com

For Further Investigation:

A simple change to the basic astable circuit will allow a 555 timer to be used as a triangle waveform generator. The circuit is shown in Figure 29-5. Construct the circuit. Vary the potentiometer and observe the results. Summarize your findings in a short report.

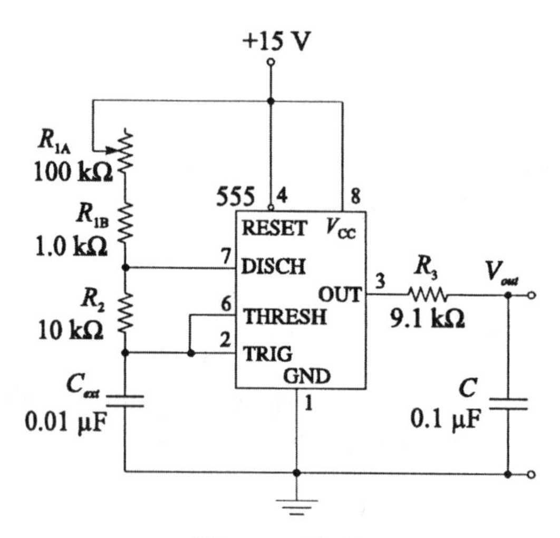

Figure 29-5

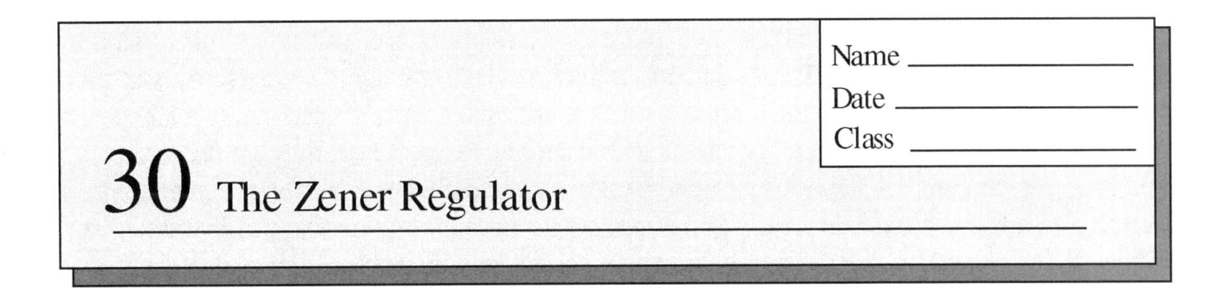

30 The Zener Regulator

Reading:
Floyd and Buchla, *Analog Fundamentals: A Systems Approach*, Sections 11-1 and 11-2

Objectives:
After performing this experiment, you will be able to:
1. Use an oscilloscope to plot the characteristic curve of a zener diode.
2. Test a zener regulator circuit for the effect of a changing source and a changing load.
3. From measurements, compute the line and load regulation of a zener regulator circuit.

Summary of Theory:
When a sufficiently large reverse bias voltage is applied to a zener diode, the reverse current will suddenly increase, as illustrated in the characteristic curve in Figure 30-1. This sudden increase happens at a voltage called the zener voltage, V_Z. A zener diode is a special diode designed to operate in this breakdown region. The schematic symbol for a zener diode is shown in Figure 30-2.

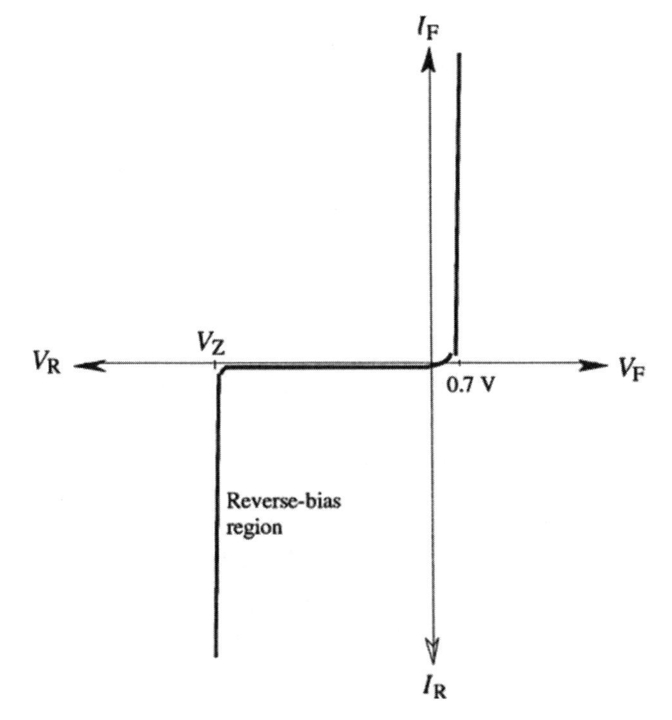

Figure 30-1 Zener characteristic.

Figure 30-2 Zener symbol.

The zener voltage is a precise voltage that varies according to the type of zener; typically it is a few volts but can be as much as several hundred volts. Although zeners are temperature sensitive, devices have been designed that can compensate for this sensitivity. Zeners are used in applications that require a constant voltage such as voltage regulators and in certain meters where they are used as a reference voltage for comparison. Ideally, the zener breakdown characteristic is a straight vertical line, but in practice, a small ac resistance is present, similar to the ac resistance in the forward-biased diode of Experiment 2. The ac resistance is found by dividing a *change* in voltage by a *change* in current measured in the vertical breakdown region. The ac resistance is typically from 10 Ω to 100 Ω.

In this experiment, you will measure a zener diode's *I-V* characteristic, then use the zener in two regulator circuits. In the first circuit, you will test the effect of a varying voltage and in the second circuit, you will test the effect of a varying load.

Materials Needed:

Resistors: one 220 Ω (1/2 W), one 1.0 kΩ, one 2.2 kΩ
One 1.0 kΩ potentiometer
One center-tapped transformer with fused primary, 12.6 V ac
One 5 V zener, (1N4733 or equivalent)

For Further Investigation:

Second 5 V zener, (1N4733 or equivalent)

Procedure:

1. Measure and record the values of the resistors listed in Table 30-1.

Table 30-1

Resistor	Listed Value	Measured Value
R_1	220 Ω	
R_2	1.0 kΩ	
R_L	2.2 kΩ	

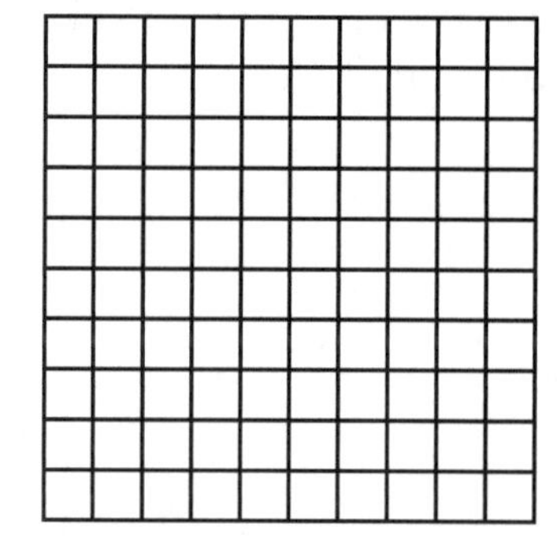

Plot 30-1

2. Observe the zener characteristic curve by setting up the circuit shown in Figure 30-3. Put scope in the X-Y mode. Sketch the *I-V* curve in Plot 30-1. The 1.0 kΩ resistor changes the scope's *y*-axis into a current axis (1 mA per volt). Label your plot for current and voltage.

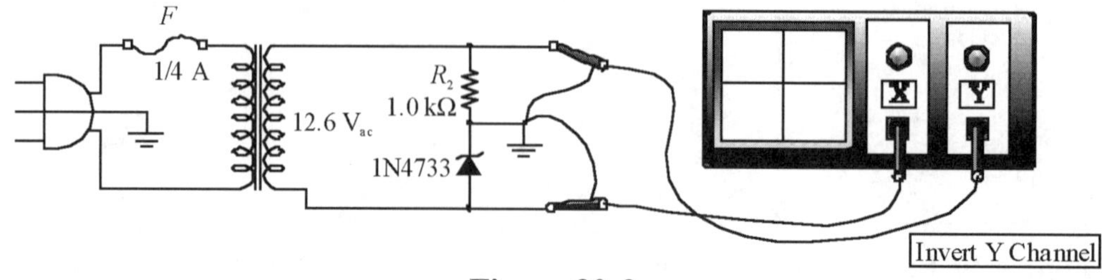

Figure 30-3

3. A common application of zener diodes is in regulators. In this step, you will investigate a zener regulator as the source voltage is varied. Connect the circuit shown in Figure 30-4. Set V_S to each voltage listed in Table 30-2 and measure the output (load) voltage, V_{OUT}.

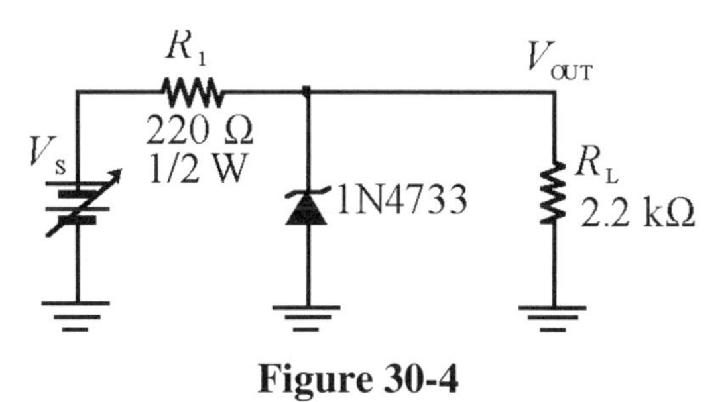

Figure 30-4

Table 30-2

V_S	V_{OUT} (measured)	I_L (computed)	V_{R1} (computed)	I_S (computed)	I_Z (computed)
2.0 V					
4.0 V					
6.0 V					
8.0 V					
10.0 V					

4. From the measurements in step 3, complete Table 30-2. Apply Ohm's law to compute the load current, I_L, for each setting of the source voltage. The voltage across R_1 (V_{R1}) can be found by applying Kirchhoff's Voltage Law (KVL) to the outside loop. It is the difference between the source voltage, V_S, and the output voltage, V_{OUT}. Note that I_s is through R_1 and can be found using Ohm's law. Find the zener current, I_Z, by applying Kirchhoff's Current Law (KCL) to the junction at the top of the zener diode.

   What happens to the zener current after the breakdown voltage is reached?

   ____

5. In this step, you will test the effect of a zener regulator working with a fixed source voltage with a variable load resistance. Often, the load is an active circuit (such as a logic circuit) in which the current changes because of varying conditions. We will simulate this behavior with a potentiometer. Construct the circuit shown in Figure 30-5. Set the power supply to a fixed +12.0 V output and adjust the potentiometer (R_L) for maximum resistance.

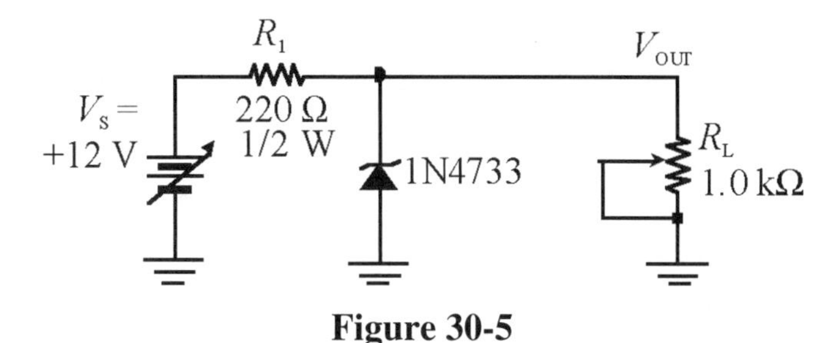

Figure 30-5

6. With the potentiometer set to 1.0 kΩ (maximum), measure the load voltage (V_{OUT}) and record the voltage in Table 30-3. Compute the other parameters listed on the first row as before. (Use Ohm's law for I_L, KVL for V_{R1}, Ohm's law for I_S, and KCL for I_Z).

Table 30-3

R_L	V_{OUT} (measured)	I_L (computed)	V_{R1} (computed)	I_S (computed)	I_Z (computed)
1.0 kΩ					
750 Ω					
500 Ω					
250 Ω					
100 Ω					

7. Set the potentiometer to each value listed in Table 30-3 and repeat step 6.

8. From the data in Table 30-3, plot the output voltage as a function of load resistance in Plot 30-2. Choose a reasonable scale factor for each axis and add labels to the plot.

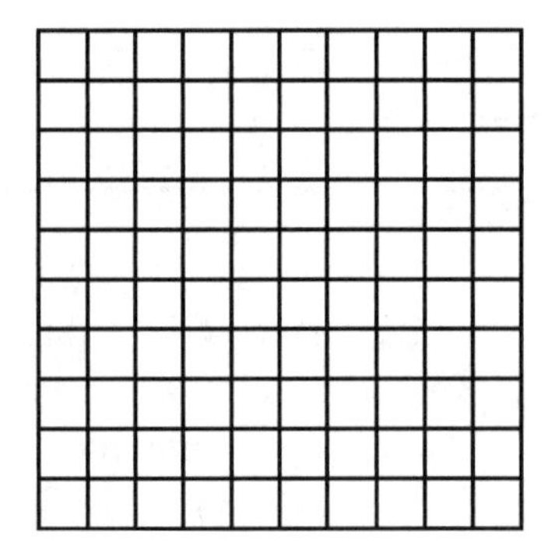

Plot 30-2

From your results, what is the smallest load resistor that can be used and still maintain regulation?

____

Conclusion:

Evaluation and Review Questions:

1. Observe the characteristic curve for a zener in Plot 30-1.
   (a) What portion of the curve is approximated by an open circuit?

   (b) What portion of the curve is approximated by a short circuit?

2. From the data in Table 30-2, compute the ac resistance of the zener when the source voltage changed from 8.0 V to 10.0 V.

3. Line regulation of a zener regulator is normally expressed as a percentage and is given by the equation:

$$\text{Line regulation} = \frac{\Delta V_{out}}{\Delta V_{in}} \times 100\%$$

   Compute the line regulation expressed as a percentage for the circuit in Figure 30-4 using the data for the *last* two rows of Table 30-2. (Note that V_{IN} in the equation is equivalent to V_S in the table.)

4. Load regulation of a zener regulator, expressed as a percentage, is given by the equation:

$$\text{Load regulation} = \frac{V_{NL} - V_{FL}}{V_{FL}} \times 100\%$$

   Compute the load regulation for the circuit in Figure 30-5. (Assume V_{OUT} for the 1.0 kΩ resistor = V_{NL} and V_{OUT} for the 100 Ω resistor represents V_{FL}).

5. Assume the potentiometer in Figure 30-5 is set to its maximum value (1.0 kΩ). Predict the output voltage for each of the following faults.

Fault	V_{OUT}
1. Zener diode is open	
2. V_S is +15 V	
3. Zener is reversed	
4. R_1 is 2.2 kΩ	
5. R_L is open	

For Further Investigation:
Modify the regulator circuit tested in this experiment by adding a second zener diode as shown in Figure 30-6 and changing the source to a 15 V_{pp} sine wave at 1.0 kHz. Verify that the source has no dc offset. Set the potentiometer to its maximum resistance. Sketch the output waveform and label voltage and time on your sketch.

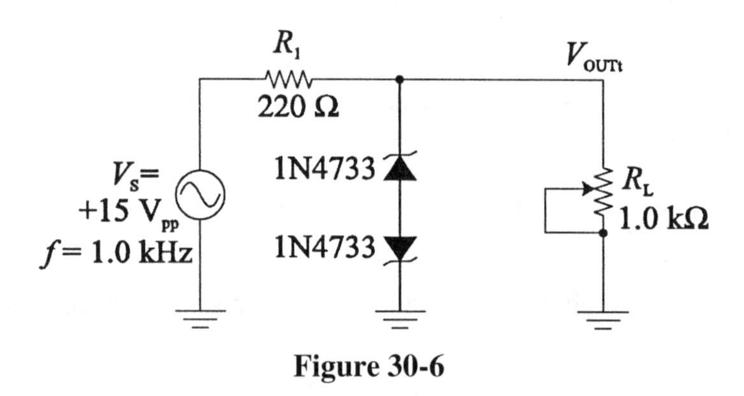

Figure 30-6

Reading:
Floyd and Buchla, *Analog Fundamentals: A Systems Approach*, Sections 11-4 through 11-6

Objectives:
After performing this experiment, you will be able to:

1. Construct and test a voltage regulator circuit. From measured voltages, compute the line and load regulation for the circuit.
2. Connect a bridge rectifier and filter to the regulator circuit and measure the input and output ripple voltage.

Summary of Theory:
Virtually all active electronic devices require a stable source of voltage. Most regulated power supplies are designed to provide a constant dc voltage to various loads from a source of alternating current (ac). Good regulation requires that the output voltage be maintained for variation in line voltages, load resistance, or temperature change. The zener regulator studied in Experiment 30 cannot provide sufficient current for most applications; however, it can form the basis of a stable reference voltage for a series regulator. The basic series regulator circuit places the pass transistor in series with the output; hence the term series regulator. The basic circuit was introduced years ago and is still one of the most popular forms of regulators.

For smaller power supplies, three-terminal IC regulators provide a stable output over a wide range of loads (see Experiment 3 – For Further Investigation). These IC regulators are typically within 5% of the nominal voltage – they are available with a limited number of fixed outputs ranging from 5 V to 24 V and can provide up to 1 A output. The 78xx series (positive regulators) and 79xx series (negative regulators) use the last two digits (xx) to represent the output voltage – for example, the 7805 is a +5.0 V regulator. The 7805 regulator can supply 1.0 A with 80 dB of ripple rejection. Integrated circuit regulators can be extended to higher currents by adding external pass transistors. Adjustable regulators allow the power supply to be varied over a wide range of values with a minimum of external components. Adjustable regulators are available with extra features such as adjustable short circuit current limiting. An example of an adjustable IC regulator is the popular LM317, adjustable from 1.2 V to 37 V.

In this experiment, you will test a series regulator constructed from an op-amp, zener reference and a series pass transistor. You will measure the line regulation, load regulation, and ripple voltage. In the For Further Investigation section, a three-terminal regulator used as a current source will be tested.

Materials Needed:
Resistors: four 330 Ω resistors, one 1.0 kΩ, one 1.2 kΩ, one 2.7 kΩ
One 1.0 kΩ potentiometer
Transistors: one 2N3904 *npn* transistor, one SK3024 *npn* power transistor
Diodes: one 1N4733 5 V zener diode, four 1N4001 rectifier diodes
One 1000 μF capacitor

For Further Investigation:

One 7805 or 78L05 regulator
One 220 Ω resistor
One LED
One 0 – 50 mA ammeter

Procedure:

1. Measure and record the values of the resistors listed in Table 31-1.

Table 31-1

Resistor	Listed Value	Measured Value
R_1	2.7 kΩ	
R_2	330 Ω	
R_3	1.0 kΩ	
R_4*	1.0 kΩ	
R_5	1.2 kΩ	
R_L	330 Ω	

*potentiometer; record maximum resistance

2. Construct the series regulator circuit shown in Figure 31-1. This circuit illustrates the concept of voltage regulation, but it is limited to relatively small power levels unless heat sinking of the pass transistor is provided. (**Caution**: the power dissipated in the pass transistor for this experiment will cause it to become hot, but a heat sink is not required for the loads specified.) Connect the input to a dc power supply set to +18.0 V.

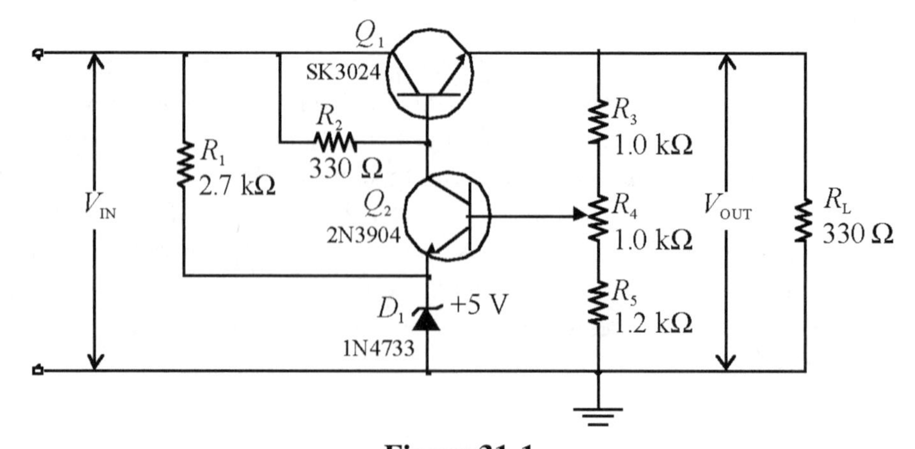

Figure 31-1

3. Compute the minimum and maximum output voltage for the regulator. The minimum voltage is found by assuming the $V_{BASE(Q2)}$ is 5.7 V (the zener drop + 0.7 V). This voltage is dropped across R_4 and R_5. Since R_3, R_4, and R_5 are in series, the output can be found by the proportion:

$$\frac{V_{OUT(min)}}{V_{BASE(Q2)}} = \frac{R_3 + R_4 + R_5}{R_4 + R_5}$$

Enter the computed $V_{OUT(min)}$ in Table 31-2. Set up a similar proportion to find $V_{OUT(max)}$. Enter the computed values in Table 31-2.

Table 31-2

Parameter	Computed Value	Measured Value
$V_{OUT(min)}$		
$V_{OUT(max)}$		

4. Maintain the input voltage at +18.0 V. Test the minimum and maximum output voltage by varying R_4 over its range. Enter the measured values in Table 31-2.

5. In this step and step 6, you will measure the variation on the output voltage due to a change in the input voltage and compute the line regulation. With the input voltage at +18 V, adjust R_4 for an output of +10.0 V. (This is shown as the starting value in Table 31-3). Then set the input voltage to each value listed in Table 31-3 and measure the output voltage. (Note that a 4 V variation on the input is more than would be observed in operation).

Table 31-3

V_{IN}	V_{OUT} (measured)
+18.0 V	+10.0 V
+17.0 V	
+16.0 V	
+15.0 V	
+14.0 V	

Table 31-4

Step	Quantity	Measured Value
6	Line regulation	
7	V_{NL}	+10.0 V
	V_{FL}	
	Load regulation	
8	$V_{ripple(in)}$	
	$V_{ripple(out)}$	

6. Compute the line regulation from the data taken in step 5. The line regulation (expressed as percent line regulation per volt) is given by Equation 11-2 in the text and repeated here for convenience:

$$\text{Line Regulation} = \left(\frac{\Delta V_{OUT} / V_{OUT}}{\Delta V_{IN}}\right) 100\%$$

Use the first and last entry of Table 31-3 to compute the line regulation and enter the value in the first row of Table 31-4.

7. The load regulation is found by determining the change in the output voltage between no load and full-load and dividing by the output voltage at full load. It is usually expressed as a percentage. For the purpose of this experiment, we will assume the full-load output current is 90 mA. Remove the 330 Ω load resistor and set the input voltage to +16 V. Adjust the output voltage to +10.0 V (V_{NL}). Then install three parallel 330 Ω resistors across the output and measure the output load voltage (V_{FL}). Compute the load regulation from the equation:

$$\text{Load Regulation} = \left(\frac{V_{NL} - V_{FL}}{V_{FL}}\right) 100\%$$

Enter the measured full-load voltage and the computed load regulation in Table 31-4.

8. In this step, you will determine ripple at the input and output of the regulator. Disconnect the power supply that you have been using as an input device and add the bridge rectifier circuit shown in Figure 31-2. (The bridge rectifier is the same circuit you studied in Experiment 3 plus a filter capacitor). The load consists of the three 330 Ω resistors in parallel from step 7. Measure the peak-to-peak ripple voltage across C_1 ($V_{ripple(in)}$) and across the output load ($V_{ripple(out)}$). Couple your oscilloscope with ac coupling to view the ripple. Record the results in Table 31-4.

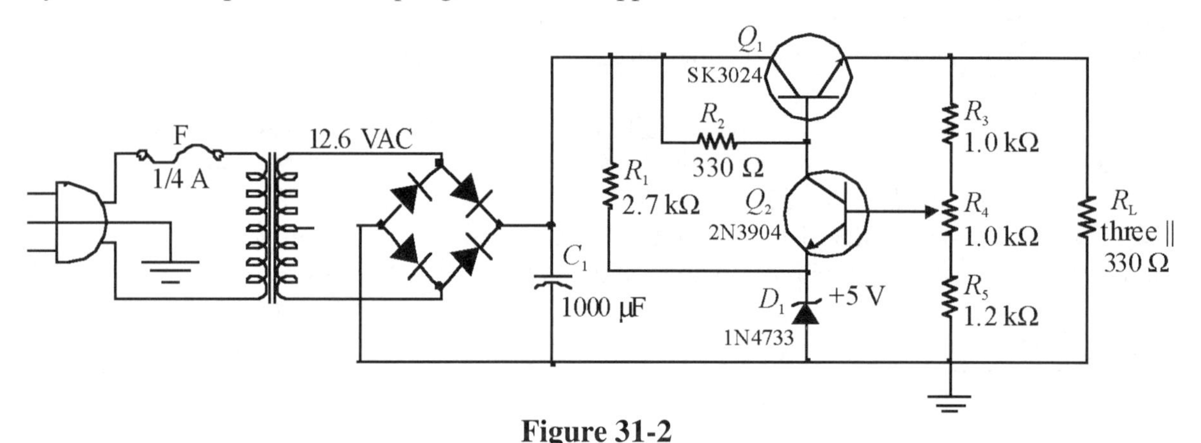

Figure 31-2

Conclusion:

Evaluation and Review Questions:

1. Assume the regulator in Figure 31-1 is designed for a maximum load current of 90 mA at 10 V and the ß of Q_1 is 100. If the input voltage is +18 V,
   (a) What is the power dissipated in the series pass transistor?

   (b) What is the base current in the series pass transistor?

   (c) What is the collector current in Q_2?

2. Assume a student wanted to supply more current than the rated 1 A for the series pass transistor by placing a second identical pass transistor in parallel with the first.
(a) What problem does this create?

(b) Why isn't it a problem if MOSFET transistors were used?

3. What change would you suggest to the circuit in Figure 31-2 if you needed to reduce the ripple?

4. The major drawback to a series regulated supply is inefficiency. The efficiency is defined as the percentage of input power that can be delivered to the load. For the circuit in Figure 31-2, assume the input power is 2.01 W (17.5 V at 115 mA). With the output set to 10 V and a 110 Ω load resistance, compute the efficiency of the regulator.

5. Figure 31-3 shows the series regulator from this experiment with the addition of current limiting (Q_3 and R_6). Assume you wanted to set the current limit to 200 mA. Select a value for R_6 that will produce this limit. Show how you obtained the value selected.

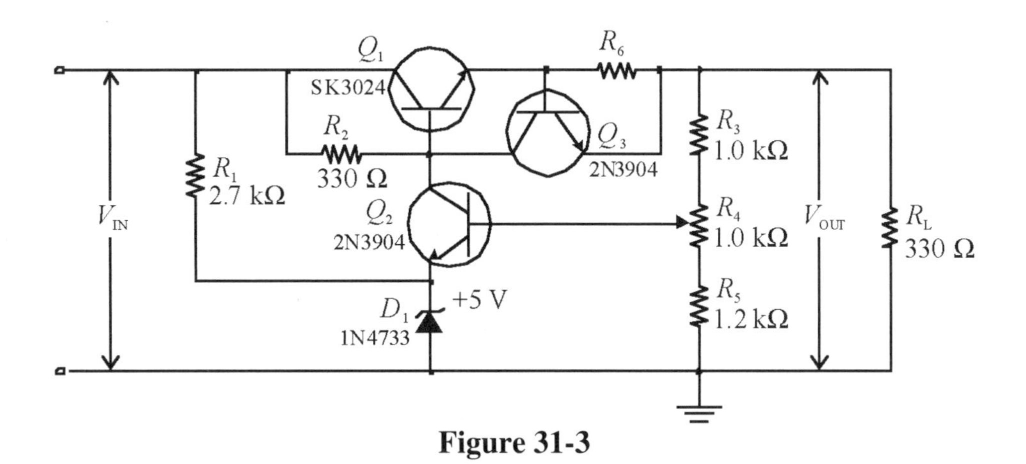

Figure 31-3

For Further Investigation:

A three-terminal regulator can be used for a current source by "programming" a current as shown in the circuit in Figure 31-4. Use either a 7805 or 78L05 regulator. The current that is programmed is limited only by the current and power rating of the regulator. Test the circuit by varying the input voltage between +8 V and +18 V and observe the current in the LED. Measure the output current as a function of the input voltage. What current is "programmed" in this circuit? How can you change this current?

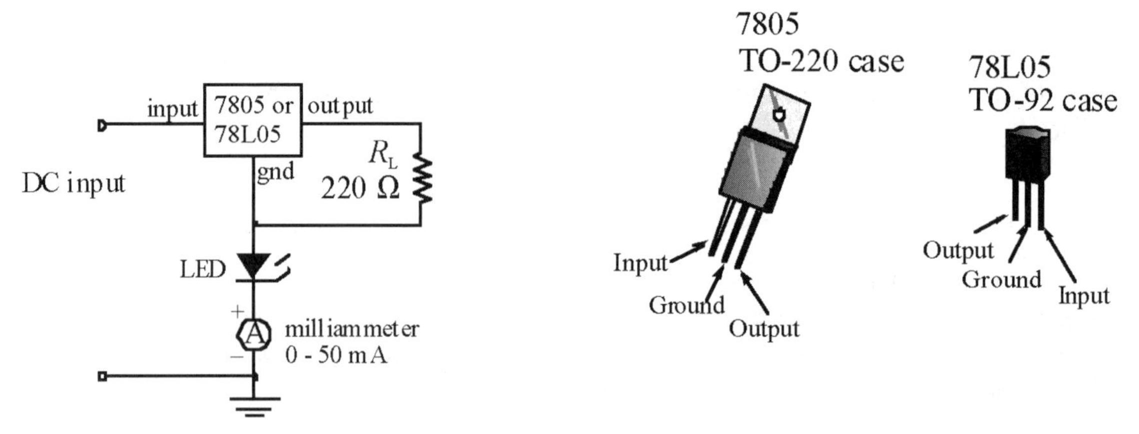

Figure 31-4

Name ______________
Date ______________
Class ______________

32 The Instrumentation Amplifier

Reading:
Floyd and Buchla, *Analog Fundamentals: A Systems Approach*, Sections 12-1 and 10-5 (for 555 timer)

Objectives:
After performing this experiment, you will be able to:
1. Construct an instrumentation amplifier (IA) using three op-amps. Compute the differential gain and measure the common-mode gain. Compute the CMRR' for the IA.
2. Construct an oscillator as an input signal source for the IA constructed in objective 1. Add simulated common-mode noise to the input and adjust the common-mode gain of the amplifier for minimum noise at the output. Demonstrate how an IA can selectively amplify a small differential signal in the presence of a large common-mode signal.

Summary of Theory:
Frequently, instrumentation systems have low-level signals from transducers (strain gages, thermistors, etc.). These signals can be extremely small and may originate from high impedance sources located some distance from the signal conditioning. Noise is frequently a problem with such signals. The requirement for accurate amplifiers for this type of application led to the development of the *instrumentation amplifier*, or IA. Integrated circuit IAs are high-performance, high-impedance amplifiers that use a differential input. Gain is normally set with a single external resistor.

An instrumentation amplifier must also have excellent dc performance characteristics – ideally it should have no dc offset or voltage drift and should reject common-mode noise. Common-mode noise is unwanted voltage that is present on both signal leads and a common reference. Frequently, electronic systems have noise that originates from a common-mode source (typically interference). In order to reject common noise, the differential gain should be high while common-mode gain should be very low (implying a high CMRR). Although most IAs are in IC form, it is possible to construct an IA using conventional op-amps by adding two op-amps to a basic differential amplifier. For this experiment, you will construct an IA using three op-amps in the configuration shown in Figure 32-1. Differential and common-mode inputs are shown to clarify their meaning. In this configuration, $R_1 = R_2$ and the differential gain is given by the equation:

$$A_{v\,(\mathrm{d})} = 1 + \frac{2\,R_1}{R_\mathrm{G}}$$

Although the resulting IA is not ideal, the results are surprising good.

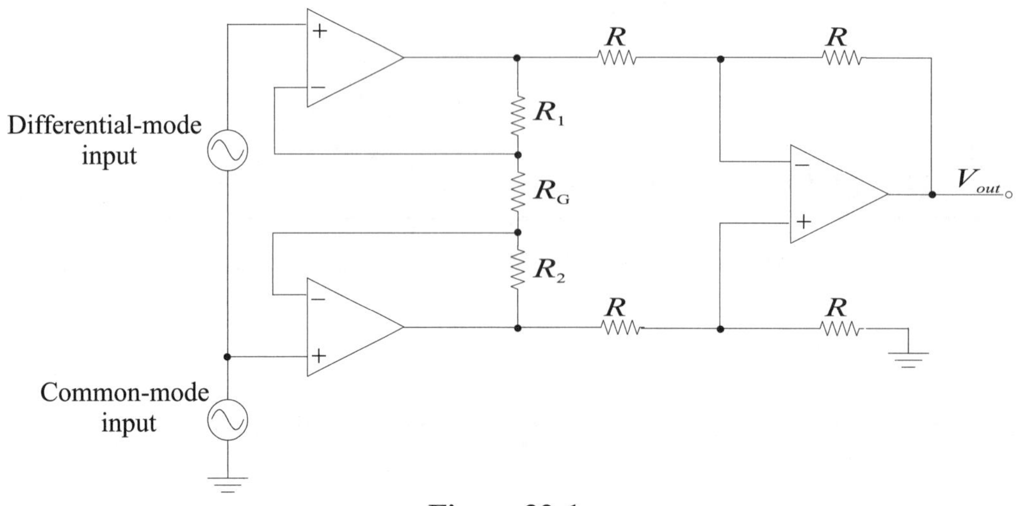

Figure 32-1

Materials Needed:
Resistors: one 22 Ω, one 470 Ω, two 1.0 kΩ, one 8.2 kΩ, five 10 kΩ, three 100 kΩ
Capacitors: one 0.01 μF, two 1.0 μF
Three LM741C op-amps
One 555 timer
One small 9 V battery
One 5 kΩ potentiometer
30 cm twisted-pair wire

For Further Investigation:

One CdS cell – Electronix Express 08GL7516 or equivalent

Procedure:
Instrumentation Amplifier Measurements

1. Measure and record the values of the resistors listed in Table 32-1. In step 3, you will use the measured values of R_1, and R_G for computing the differential gain. For best results, R_1 and R_2 should match.

Table 32-1

Resistor	Listed Value	Measured Value
R_1	10 kΩ	
R_2	10 kΩ	
R_G	470 Ω	
R_3	10 kΩ	
R_4	10 kΩ	
R_5	10 kΩ	
R_6	8.2 kΩ	
R_8	100 kΩ	
R_9	100 kΩ	

2. Construct the circuit shown in Figure 32-2. The instrumentation amplifier (IA) is shown in the dotted box. It is driven by the generator in differential mode. The purpose of R_8 and R_9 is to assure a bias path for the op-amps when the source is isolated in step 6. To simplify the schematic, the power supply connections and bypass capacitors are not shown. (Use ±15 V for all op-amps; use two 1.0 μF bypass capacitors installed near one of the op-amps as in earlier experiments.)

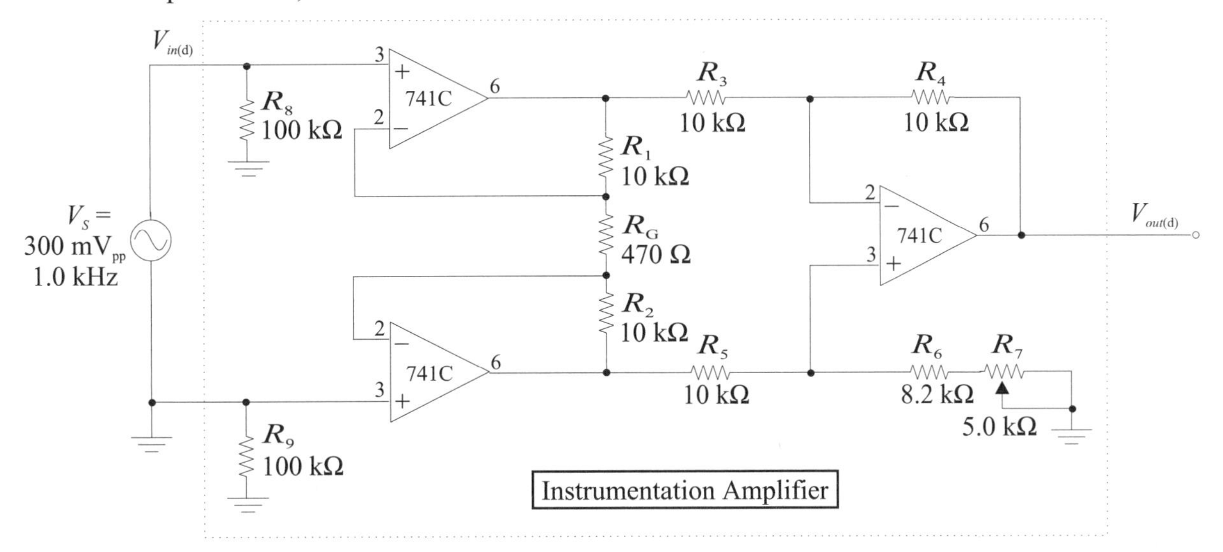

Figure 32-2 Circuit to measure differential parameters.

3. Set the potentiometer (R_7) in the middle of its range. ($R_6 + R_7$ should add to approximately 10 kΩ.) Set the input for a 300 mV$_{pp}$ sine wave at 1.0 kHz. This represents the differential-mode input signal, $V_{in(d)}$. Compute the differential gain from the equation given in the Summary of Theory. Using the computed gain, compute the expected differential output voltage $V_{out(d)}$. Then measure these parameters and record the measured values in Table 32-2.

Table 32-2

Step	Parameter	Computed Value	Measured Value
3	Differential Input Voltage, $V_{in(d)}$	300 mV$_{pp}$	
	Differential Gain, $A_{v(d)}$		
	Differential Output Voltage, $V_{out(d)}$		
4	Common-mode Input Voltage, $V_{in(cm)}$	10 V$_{pp}$	
	Common-mode Gain, $A_{v(cm)}$		
	Common-mode Output Voltage, $V_{out(cm)}$		
5	CMRR′		

4. Drive the IA with a common-mode signal as shown in Figure 32-3. Set the signal generator for a 10 V_{pp} signal at 1.0 kHz ($V_{in(cm)}$) and measure this input signal. Observe the common-mode output voltage and adjust R_7 for *minimum* output. Measure the peak-to-peak output voltage, $V_{out(cm)}$. Determine the common-mode gain, $A_{v(cm)}$, by dividing the measured $V_{out(cm)}$ by the measured $V_{in(cm)}$. Record all values in Table 32-2.

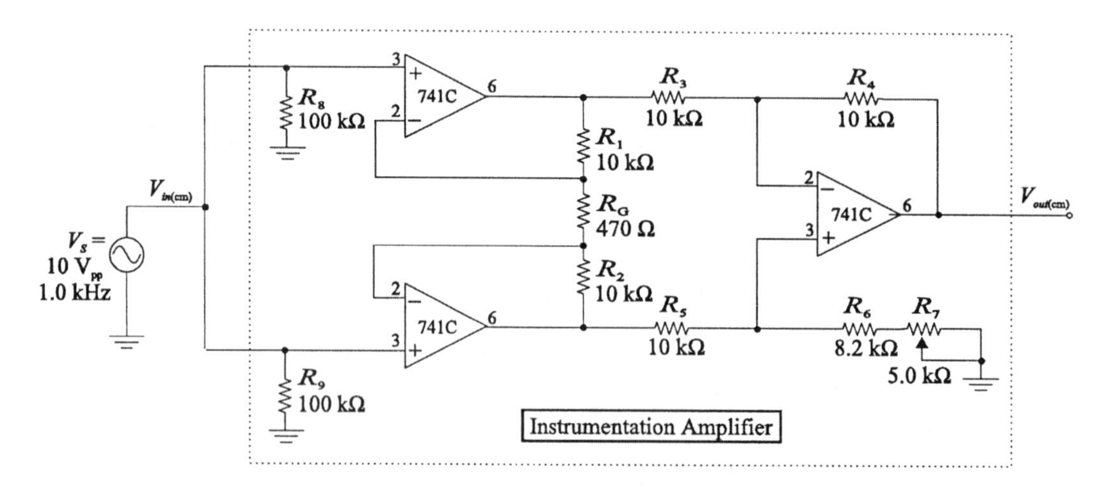

Figure 32-3 Circuit to measure common-mode parameters.

5. Determine the CMRR′ (in dB) from the equation $\text{CMRR}' = 20\ \log \dfrac{A_{v\,(d)}}{A_{v\,(cm)}}$. Enter this as the measured value in Table 32-2. When you have completed the common-mode measurements, reduce the signal generator frequency to 60 Hz. This will represent the noise source for the last part of the experiment.

Adding Differential-mode and Common-mode Sources

6. In this step, you will build a pulse oscillator to serve as a differential signal source for the instrumentation amplifier. The oscillator is a 555 timer (discussed in Section 17-6 of the text). Construct the circuit shown in Figure 32-4 (preferably on a separate protoboard if you have one available). R_C and R_D serve as an output voltage divider to simulate a small signal source (such as a transducer). Measure the output frequency and voltage and indicate these values in the first two rows of Table 32-3.

   Note that the differential signal source must be *floating* (no common ground with IA) so it is powered by a small 9 V battery as shown.

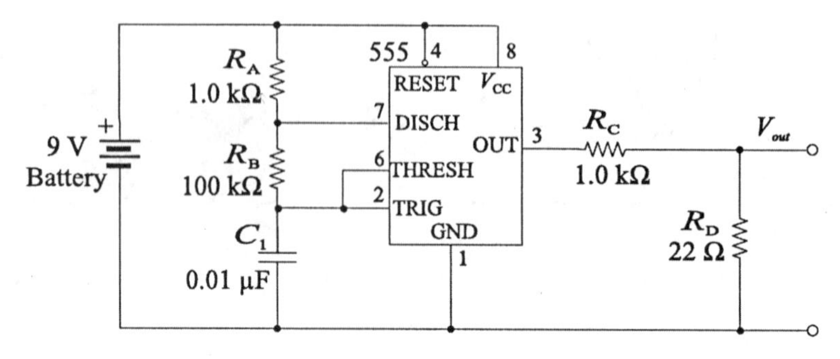

Figure 32-4 Oscillator to serve as a source for the instrumentation amplifier.

Table 32-3

Parameter	Measured Value
Oscillator frequency	
$V_{out(pp)}$ from oscillator	
$V_{out(pp)}$ from IA	

7. Connect the oscillator to the IA as shown in Figure 32-5 with about 30 cm of twisted-pair wire to simulate a short transmission line. Be sure there is no common ground from the oscillator to the input of the instrumentation amplifier. Measure the output signal from the IA and record the output in the last row of Table 32-3. Note that this signal represents a differential-mode input and is therefore amplified by the differential gain, $A_{v(d)}$.

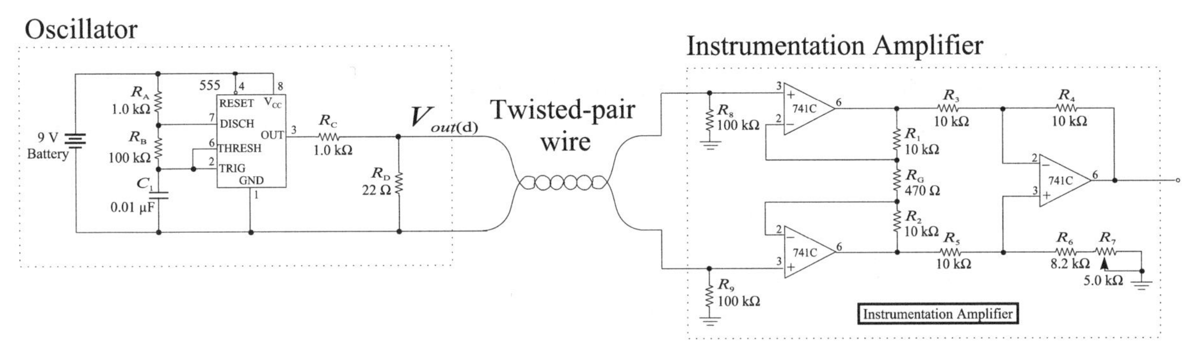

Figure 32-5 Connecting the oscillator to the instrumentation amplifier.

8. Now you will add a simulated source of common-mode noise to the oscillator. Often, the noise source is 60 Hz power line interference, but seldom is it as large or well connected as in this step. Set up your function generator for a 10 V_{pp} sine wave at 60 Hz to simulate a large amount of common-mode interference from a power line. Connect the generator to one side of the oscillator as shown in Figure 32-6. Observe the output signal from the IA. Adjust R_7 for minimum common-mode signal.

   Observations:____

   ____

Oscillator

Instrumentation Amplifier

Function generator as "noise" source 60 Hz 10 V_{pp}

Figure 32-6 Adding a common-mode "noise" source to the oscillator.

Conclusion:

Evaluation and Review Questions:

1. What type of noise could *not* be rejected by an instrumentation amplifier?

2. Some instrumentation amplifiers have a CMRR' of 130 dB. Assuming the circuit in this experiment had a CMRR' of 130 dB with the same differential gain, what common-mode output would you expect?

3. Explain how the IA was able to pass the oscillator signal while simultaneously blocking the signal from the function generator.

4. Why was it important to power the 555 timer using a separate battery rather than use the same supply that powered the op-amps?

5. What advantage does an IA, such as the one that you constructed, have over an ordinary differential amplifier?

For Further Investigation:

Many transducers convert a physical quantity (light, pressure, speed) into an oscillation. The resulting frequency will depend on a capacitance or resistive change in the transducer. This type of change can easily be observed in this experiment by replacing R_B in the oscillator with a CdS cell. The resistance of a CdS cell is a function of the incident light.

Replace R_B with a CdS cell and observe the effect on the output signal from the IA as you cover the cell. Measure the lowest and highest frequency you can obtain. Can you think of an application for this circuit? Summarize your results.

Name ____________
Date ____________
Class ____________

33 Log and Antilog Amplifiers

Reading:
Floyd and Buchla, *Analog Fundamentals: A Systems Approach*, Section 12-4

Objectives:
After performing this experiment, you will be able to:
1. Construct log and antilog amplifiers and observe signal compression and reconstruction.
2. Observe and sketch the output of the log amp for a triangle and sine wave input.
3. Measure the performance of the log amplifier and plot the transfer characteristic.

Summary of Theory:
Dynamic range is defined as the difference between the smallest and largest signal of interest. Signals with a large dynamic range can present a problem for A/D conversion, transmission, and recording. For such signals, the dynamic range can be compressed with a logarithmic[1] amplifier (*log amp*). In other words, a log amp is a nonlinear amplifier that converts a large change in the input signal to a small change in the output signal. In addition to signal compression applications, log amps are also useful for certain arithmetic operations with analog quantities. The For Further Investigation has an arithmetic processing application.

Log amps exploit the logarithmic *I-V* characteristic of the *pn* junction. As the voltage is increased across the junction, the current rises exponentially (see Experiment 2 – Further Investigation). The basic log amp is formed by putting a *pn* junction in the feedback path of an inverting amplifier, causing the output voltage to be proportional to the logarithm of the input voltage. The input signal is converted to a current source by the input resistor and the virtual ground. The diode converts this current to a small voltage drop across the junction. As a result, the dynamic range of the input signal is compressed (and is opposite polarity to the input since it is an inverting amplifier).

Although a diode in the feedback path forms a basic log amp, plain diodes are not generally used because they have less dynamic range than transistors. A log amp with a transistor in the feedback path is shown in Figure 33-1(a). Notice that the virtual ground causes the base and collector to have almost exactly the same potential. The capacitor prevents oscillation that occurs due to the transistor's tendency to add gain inside the feedback loop. The diode prevents the transistor from being destroyed in the event the output tries to go positive, causing breakdown of the base-emitter diode. While this basic circuit is satisfactory to illustrate the process of dynamic range compression, it is temperature sensitive. In addition, no gain control is included in the basic circuit. More practical circuits, including all integrated circuit log amps include temperature compensation and gain control.

---

[1] A logarithm is simply an exponent. The definition of a logarithm is that it is the exponent to which a base quantity must be raised to equal a given number.

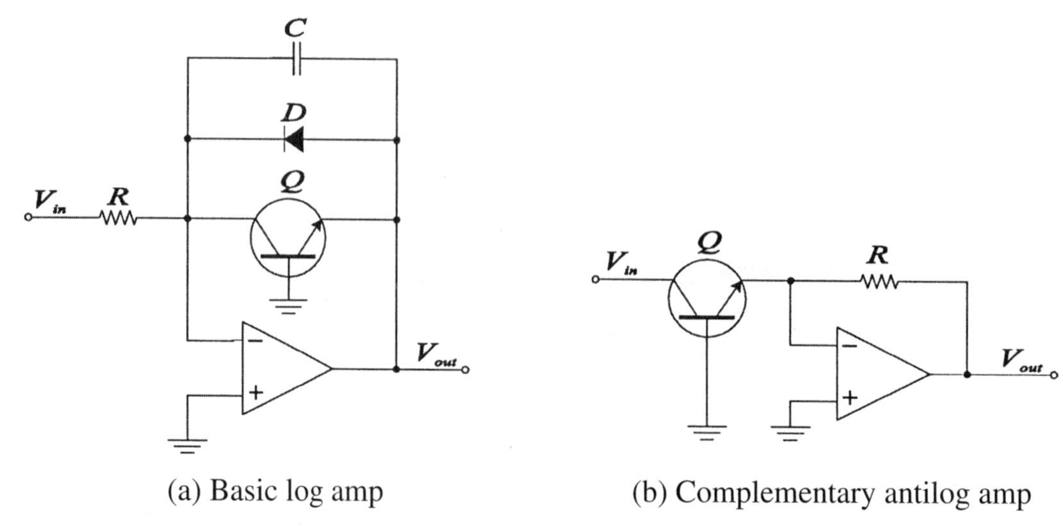

(a) Basic log amp (b) Complementary antilog amp

Figure 33-1

After signal compression, the resulting log signal is usually processed in some manner, sent over a transmission path or stored. When it is necessary to reconstruct the signal, a complementary *antilog amplifier* (also known as an *exponential amplifier*) is needed. An antilog amplifier is illustrated in Figure 33-1(b). An antilog amplifier reverses the process of signal compression and restores the signal to its original form. In this experiment, you will construct a positive-input log amp that produces a negative output. To reconstruct the signal, you will need a negative-input antilog amplifier (which produces the original positive input). Log amps constructed from low-bias current op-amps can have a dynamic range of five decades of input voltage and IC log amps can do even more. The 741C op-amps used in this experiment will not allow a high dynamic range but will serve to illustrate the concepts of signal compression and reconstruction.

Materials Needed:

Resistors: two 100 kΩ
Capacitors: one 0.01 μF, two 1.0 μF
Two 1N914 signal diodes (or equivalent)
Two 741C op-amps
Two 2N3904 *npn* transistors (or equivalent) (*Note:* You will obtain best results if ßs match).

For Further Investigation:

One additional 741C op-amp
Resistors: one 5.1 kΩ, two 10 kΩ
Potentiometers: one 500 Ω, one 10 kΩ

Procedure:

1. Measure and record the values of the resistors listed in Table 33-1.

Table 33-1

Resistor	Listed Value	Measured Value
R_1	100 kΩ	
R_2	100 kΩ	

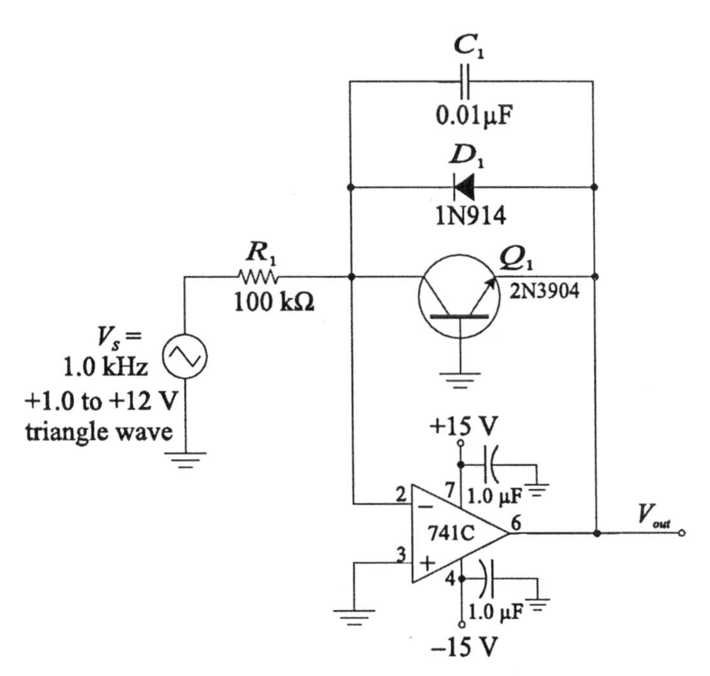

Figure 33-2

2. Construct the log amp shown in Figure 33-2. The log amp shown is designed *only* for positive signals at the input. Before connecting the function generator to the circuit, set the signal to a 1.0 kHz, +1.0 V to +12 V positive triangle waveform. Adjust the generator's dc offset control to achieve a positive waveform.

3. Observe the waveform at the output of the log amp. Sketch the input and output waveforms on Plot 33-1. Label the voltage and time on the plot. Notice the polarity of the output.

4. Change the input waveform to a positive sine wave (+1.0 V to +12 V). Sketch the input and output waveforms on Plot 33-2. Label the voltage and time on the plot.

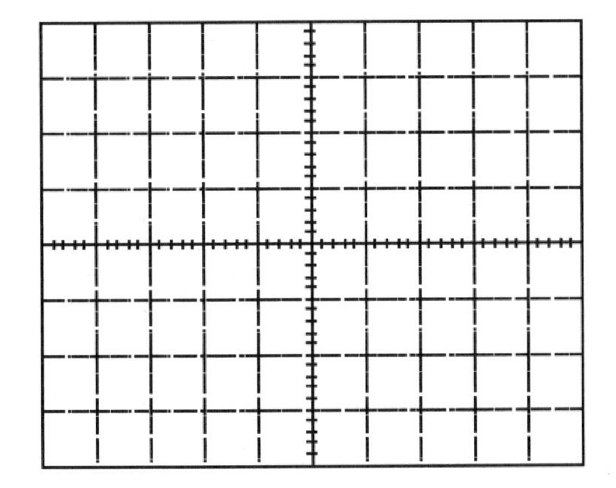

Plot 33-1

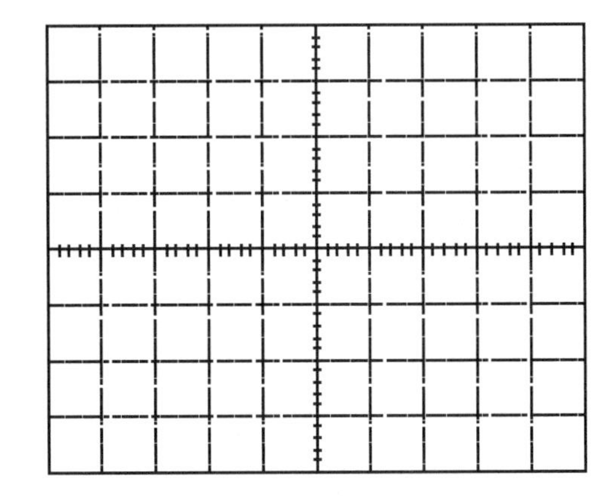

Plot 33-2

5. Connect the antilog amp shown in Figure 33-3 to the output of the log amp. The power inputs (not shown) should be connected to ±15 V. If the circuit is working correctly, you should see that the output waveform from the antilog amp matches the input to the log amp. Check both the sine wave and the triangle wave you tested earlier.

   Observations:____

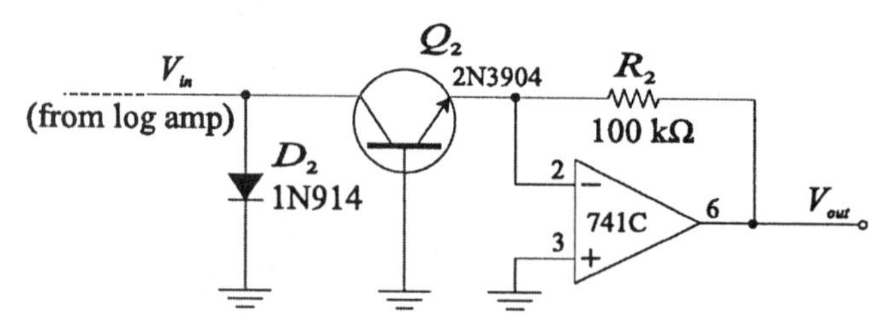

Figure 33-3

6. Check the temperature sensitivity of the circuit by grasping the transistor on the log amp between your fingers and note what happens to the output from the antilog amp. Try the same test on the diode on the antilog amp.

   Observations:____

7. To obtain quantified data on the log amp, replace the function generator with a positive dc power supply. Set the supply to each voltage listed in Table 33-2 and measure the output of the log amp using a DMM. Plot the results on the semilog plot shown in Plot 33-3.

Table 33-2 Data for Log Amp

V_{IN}	V_{OUT}
+1.0 V	
+2.0 V	
+4.0 V	
+6.0 V	
+8.0 V	
+10.0 V	
+12.0 V	

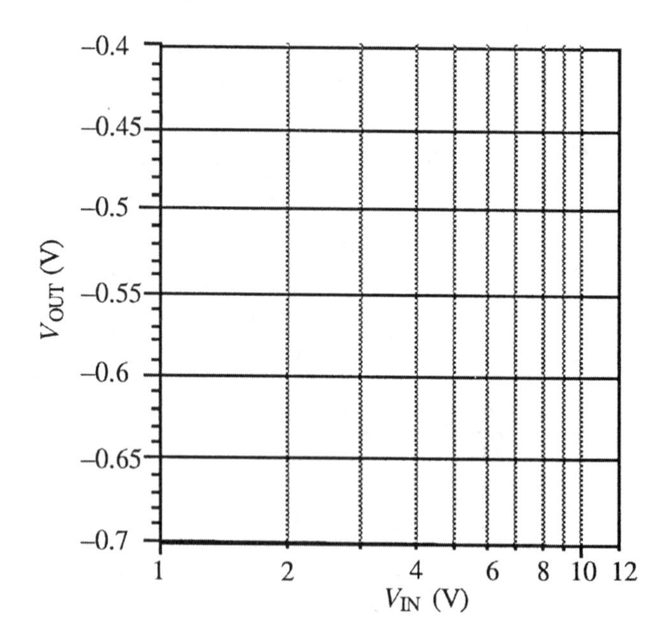

Plot 33-3

Conclusion:

Evaluation and Review Questions:

1. Cite at least three problems with the log amp in this experiment that are improved in commercial IC log amps.

2. Plot 33-3 represents the transfer curve for the log amp. (Recall that a transfer curve is the output plotted against the input.) Describe the transfer curve for the antilog amp. Should it be plotted on the same type of paper to obtain a straight line response?

3. The antilog amp in the experiment used the same value resistor as the log amp. What would happen if they were different? (Can you obtain gain this way?)

4. Figure 33-4 illustrates a basic arithmetic application for log - antilog amplifiers. Assume that all amplifiers invert their inputs and the summing amplifier has unity gain (-1).
   (a) What arithmetic operation is performed?

   (b) Assuming proper calibration, what is V_{OUT}?____________

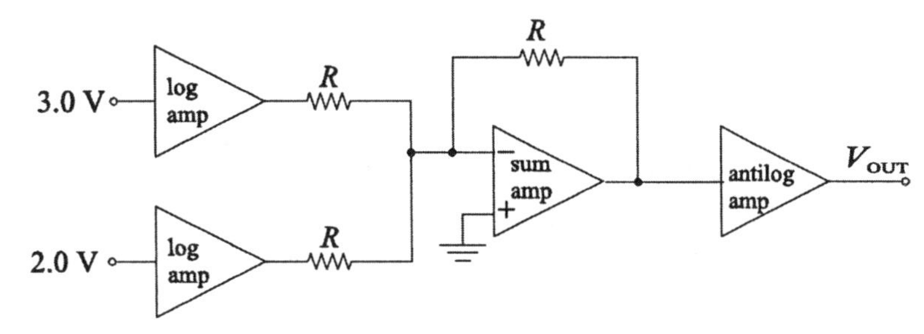

Figure 33-4

5. What simple modification could you make to Figure 33-4 to produce a circuit in which the output represents a given input voltage squared?

For Further Investigation:

Log and antilog amplifiers are readily adapted to doing exponential arithmetic operations on the input analog voltage(s). For example, multiplication of two voltages can be accomplished by finding the log of each voltage, summing the logs, and finding the antilog to return the answer. Finding the square root of a voltage is found by dividing the log by two, and finding the antilog.

The log-antilog amplifiers in this experiment are satisfactory for signal compression but are not directly suited for arithmetic operations because of an offset problem. Note that the log of 1.0 is zero (ten raised to the zero power is one). The log amp in this experiment returns a negative voltage for an input voltage of 1.0 V, which is meaningless in terms of a math operation. The negative voltage is compensated for by the antilog amplifier but the nonzero result for log 1.0 needs to be corrected.

The circuit in the dotted box in Figure 33-5 will do the job. Connect all three op-amps to ±15 V like the first one. The voltage divider consisting of R_4 and R_5 divides the log of V_{IN} by two, while R_6 (a potentiometer) compensates for the offset problem. The voltage-follower is necessary to avoid loading the divider network. Set up the circuit and adjust V_{IN} for +1.00 V using a DMM; voltages need to be set accurately with a DMM because of the very small output range of the log amp. Now adjust R_6 until the output also reads +1.00 V. The circuit is now calibrated for square roots. Test values of input from +1.00 V to +12 V. The output voltage should be equal to the square root of the input voltage. If you observe a systematic error in the output, try recalibrating using an input of +4.00 V and setting R_6 until the output reads +2.00 V. Then test other input values. Summarize your results in a short report.

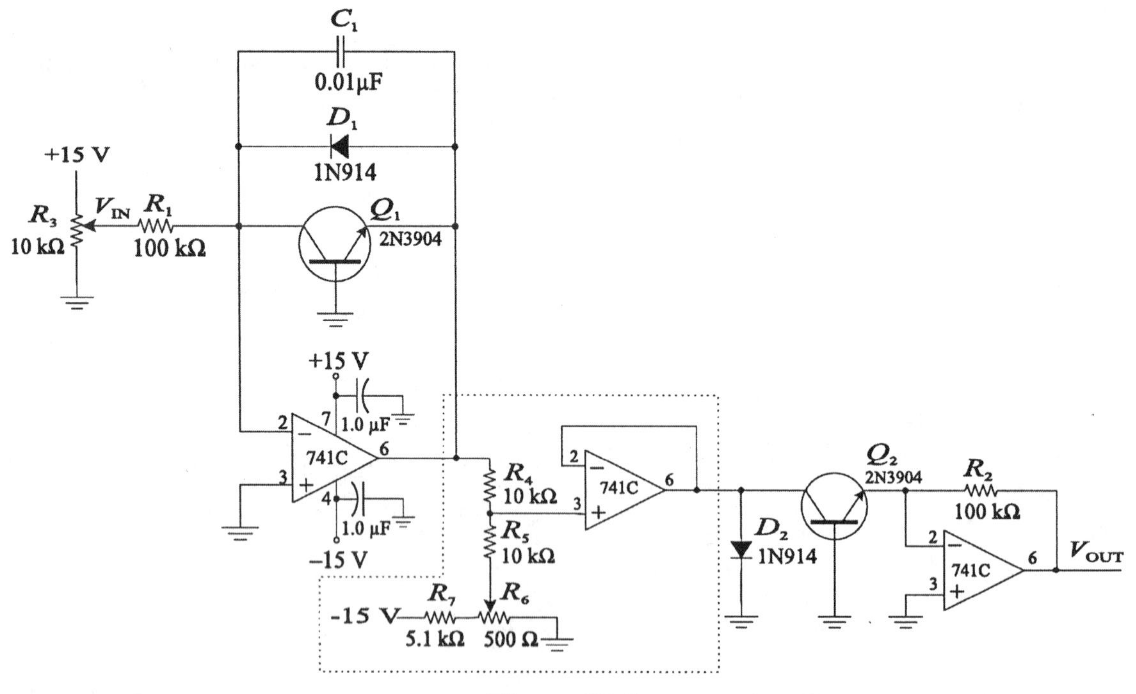

Figure 33-5

Name ______________
Date ______________
Class ______________

34 IF Amplifiers

Reading:
Floyd and Buchla, *Analog Fundamentals: A Systems Approach*, Section 13-6

Objectives:
After performing this experiment, you will be able to:
1. Construct an intermediate frequency amplifier and measure its dc and ac parameters including gain, bandwidth, and Q.
2. *For Further Investigation:* Determine the effect of different loads on the Q of the circuit.

Summary of Theory:
An IF (intermediate frequency) amplifier is an example of a tuned amplifier that is specially designed for a narrow band of frequencies. Tuned amplifiers are commonly used in high-frequency communication systems and in instruments such as the spectrum analyzer. They usually account for most of the required gain in a communication system. As explained in the text, a radio frequency (RF) is converted to a lower frequency by mixing the RF with a local oscillator. The oscillator frequency is variable, and "tracks" the incoming signal at a fixed difference frequency producing the IF which is fixed.

Both IF and RF amplifiers use similar principles of operation. Although an IF amplifier is nearly identical in principle to an RF amplifier, it differs from most RF amplifiers because it is designed to amplify only a *fixed* narrow frequency band that is not adjustable (by the operator). In this experiment, you will test an IF amplifier and observe the response of the amplifier at and near its resonant frequency.

The IF amplifier for this experiment is shown in Figure 34-1. The IF transformer in the collector circuit is a standard IF transformer that is used in many radios; it is tuned to 455 kHz, forming a narrow band, high Q circuit. The transformer has a small capacitor (about 180 pF) in parallel with the primary, forming a parallel resonant circuit. The entire assembly is mounted inside a shielded metal enclosure that in normal operation is grounded. The exact IF frequency is adjusted by using a nonmetallic screwdriver to adjust a tuning slug in the core. In communication applications, the IF gain is usually quite high for small input signals but is reduced automatically (with automatic gain control) when the input signal is large. To avoid noise problems in this experiment, the gain of this circuit has been reduced by the inclusion of a relatively large swamping resistor in the emitter circuit.

Circuit construction in high-frequency circuits should be done carefully, keeping lead lengths to a minimum and ensuring that all grounds return to a single point. To avoid noise, a bypass capacitor may be needed across the power supply.

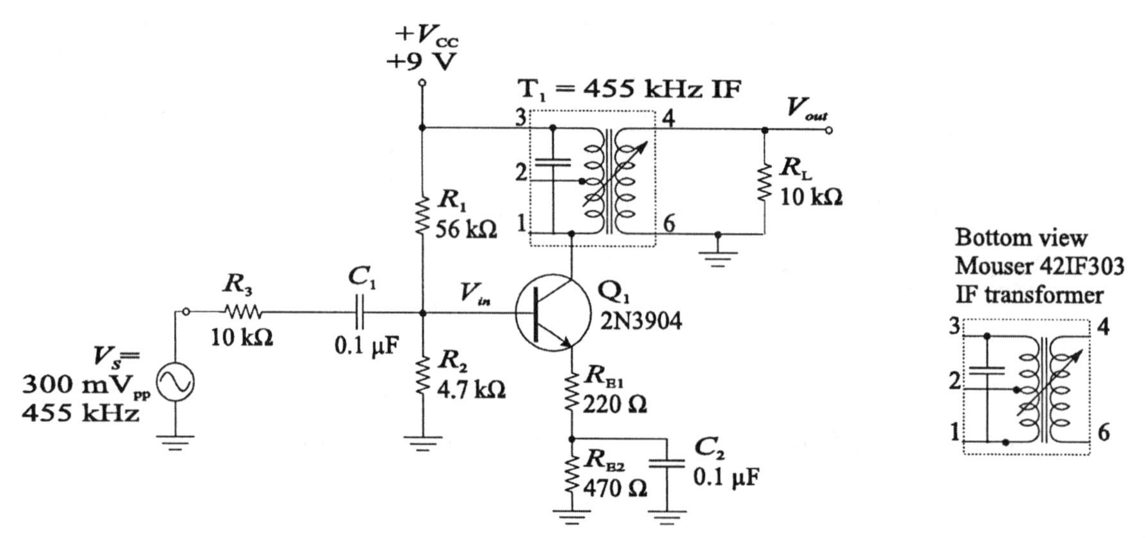

Figure 34-1 An IF amplifier.

Materials Needed:

3rd stage IF transformer (20 kΩ primary to 5 kΩ secondary) Mouser 42IF303 (or equivalent)
Resistors: one 220 Ω, one 470 Ω, one 4.7 kΩ, two 10 kΩ, one 56 kΩ
Capacitors: two 0.1 μF
Transistors: two 2N3904 npn (or equivalent)
Frequency counter (if available) (An oscilloscope can substitute.)

For Further Investigation:

Resistors: one 100 Ω, one 270 Ω, one 1.0 kΩ, one 2.0 kΩ

Procedure:

1. Measure and record the values of the resistors listed in Table 34-1.

Table 34-1

Resistor	Listed Value	Measured Value
R_1	56 kΩ	
R_2	4.7 kΩ	
R_3	10 kΩ	
R_{E1}	220 Ω	
R_{E2}	470 Ω	
R_L	10 kΩ	

Table 34-2

DC Parameter	Computed Value	Measured Value
V_B		
V_E		
I_E		
V_C		
V_{CE}		

2. Compute the dc parameters listed in Table 34-2 for the IF amplifier shown in Figure 34-1. (Review Experiment 7 for the method.) V_B, V_E, and V_C are referenced to ground. Because of the low dc resistance of the transformer primary (about 5 Ω), assume that $V_C = V_{CC}$. Enter your computed values in Table 34-2.

3. Construct the amplifier shown in Figure 34-1. The signal generator should be turned off. Measure and record the dc voltages listed in Table 34-2.

4. The input signal, V_{in}, is measured at the transistor's base. This input voltage will *increase* at resonance due to the reduced loading on the generator and will depend to some extent on the exact frequency of the resonant circuit. Set the function generator for a 300 mV_{pp} sinusoidal wave at 455 kHz. A low capacitance probe should be used for measurements to avoid loading effects. While observing the generator, adjust the frequency (slightly) for maximum amplitude (or tune the transformer for maximum amplitude). You will need to readjust the level for the 300 mV_{pp} value at the point where the frequency peaks. This is both V_{in} and the ac base voltage, V_b. Record the measured value as V_b in Table 34-3.

5. The collector voltage and the gain depend on the exact Q of the primary resonant circuit, the amount of coupling between the primary and secondary coils, the load on the secondary, and the ac resistance of the emitter circuit. In addition, loading effects from the measuring instrument can affect the results and this is why a low-capacitance probe was suggested. Measuring the primary side of the transformer will produce a greater loading effect than the secondary; however, it is necessary to view the primary to measure the gain. Measure the remaining ac parameters listed in Table 34-3. The gain, A_v, is the ratio of the ac collector voltage to the ac base voltage, V_b. The output voltage is measured at the secondary of the transformer (across the load resistor).

Table 34-3

AC Parameter	Measured Value
V_b	
V_c	
A_v	
$V_{out(\text{tot})}$	

Table 34-4

AC Parameter	Measured Value
f_c	
f_{cu}	
f_{cl}	
BW	
Q	

6. In this step, you will measure the center frequency of the IF amplifier. It is much easier to measure it accurately if you have access to a frequency counter. Like the scope, the frequency counter can load the circuit; it will have the smallest effect if used on the secondary side.

   While observing the output voltage on an oscilloscope, adjust the frequency for the maximum output. Then adjust the oscilloscope so that the peak-to-peak output voltage covers exactly 5 vertical divisions. You will probably have to take the oscilloscope out of vertical calibration to set this level exactly. Measure and record the frequency of the maximum output using a frequency counter (or the oscilloscope if a frequency counter is unavailable). Record this as the center frequency, f_c, in Table 34-4.

7. In this step, you will measure the bandwidth (BW) of the IF amplifier. Again, a frequency counter is the best way to measure the frequency accurately. Raise the generator frequency slowly while observing the output on an oscilloscope. Adjust the frequency until the peak-to-peak output voltage indicates 3.5 divisions (70%). The frequency above the center frequency at which the output is 70% of the maximum is the upper cutoff frequency (f_{cu}). Record it in Table 34-4.

8. Adjust the generator frequency for the lower cutoff frequency by watching for the 70% point below the center frequency. Record this as the lower cutoff frequency, f_{cl}, in Table 34-4.

9. Compute the *BW* of the circuit by subtracting the lower cutoff frequency from the upper cutoff frequency. Compute the *Q* of the circuit by dividing the center frequency by the *BW*. Enter the *BW* and *Q* in Table 34-4.

Conclusion:

Evaluation and Review Questions:

1. If you wanted to determine if the oscilloscope probe had a loading effect on the circuit, you could connect a second identical probe to the same point in the circuit. Explain how this would allow you to see if probe loading was a factor.

2. In Experiment 8, the value of the emitter bypass capacitor was 47 μF, yet in this experiment, it was only 0.1 μF. Explain the reason for this difference.

3. Explain why the voltage gain of the CE amplifier in this experiment was highest at the resonant frequency.

4. Assume a student determined that the voltage from the generator was 400 mV_{pp} and the drop across R_3 was 100 mV_{pp}. For this amplifier, what is the input resistance?

5. If you measured 0 Vdc on the collector, which of the following could account for the problem?
   (a) open primary
   (b) open secondary
   (c) open R_1
   (d) open R_{E1}
   (e) power supply off

For Further Investigation:

Among other things, the bandwidth of the circuit depends on the loading effects from the secondary. Does the load also affect the peak center frequency? Investigate these effects by placing a series of different loads on the secondary (in place of R_L). Suggested resistors to be used for loads are 2.0 kΩ. 1.0 kΩ, 270 Ω, and 100 Ω.

Name ______________
Date ______________
Class ______________

35 The ADC0804 Analog-to-Digital Converter

Reading:
Floyd and Buchla, *Analog Fundamentals: A Systems Approach*, Sections 14-3 through 14-6

Objectives:
After performing this experiment, you will be able to:

1. Construct an analog to digital converter circuit using an ADC0804 operating in free-running mode. Calibrate the step width and check the transfer function for a few steps.
2. Describe the basic signals in and from the ADC0804.
3. Determine the binary output pattern for a particular input voltage.
4. Interface an LM335 Kelvin temperature sensor to an ADC0804 and calibrate it to read Centigrade temperatures.

Summary of Theory:
Most physical quantities originate in analog (continuous) form. These quantities can be converted into an electrical parameter (typically voltage) by a transducer. Before processing with a digital computer, it is necessary to convert the electrical analog quantity to a number. An analog-to-digital converter (ADC) is a circuit that performs this conversion.

Conversion is a sampling process that produces an error between the actual input and the digitized representation. Several factors, discussed in the text, contribute to this error. Two critical considerations in choosing a particular ADC are the amount of error that can be tolerated in the digitized result and the conversion speed. Other considerations include the range of the input voltage, the format of the output, and power supply requirements.

The ADC0804 is a widely used 8-bit successive approximation type with parallel outputs that can be latched or put into a high impedance state for direct interfacing with a computer data bus. The ADC0804 operates on a nominal +5.0 V supply voltage. Frequently, the analog voltage of interest is less than 0 V to +5.0 V so the *span* (actual range of the input voltage) needs to be reduced. This is accomplished by a separate dc input to the ADC0804. In addition to a variable range, the ADC0804 can also accommodate different full-scale input voltages.

The sequence for the conversion process is shown in four steps in Figure 35-1. First, the internal latches are reset by making both the $\overline{WR}$ (write) input and the $\overline{CS}$ (chip select) input lines low. Conversion begins within 1 to 8 clock cycles after $\overline{WR}$ or $\overline{CS}$ makes a low to high (0 to 1) transition. Because the ADC0804 is a successive approximation ADC, it requires a clock. An internal clock can be selected that is configured by an external resistor and a capacitor or an external system clock can be used. After a maximum of 64 clock cycles, conversion is complete, signaled by the $\overline{INT}$ (interrupt) line from the ADC0804 going low. If the data is sent to a computer, the computer checks the

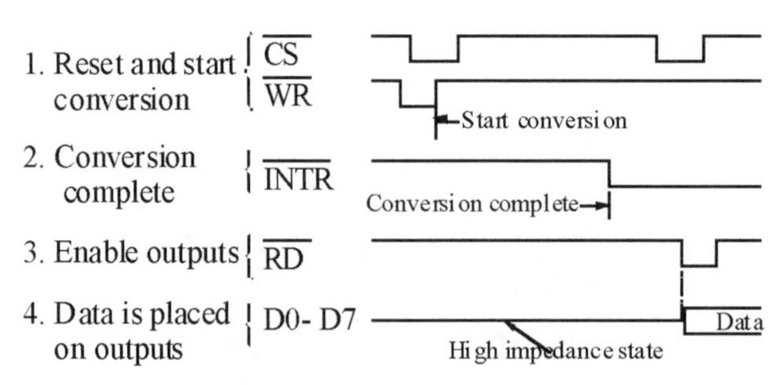

Figure 35-1 Timing for the ADC0804.

the $\overline{INT}$ line for this low and, if ready, issues a $\overline{CS}$ followed by a $\overline{RD}$ (read) input. These two signals are required to be low to enable the output latches and place the data on the data bus.

The ADC0804 can operate in a free-running mode. In the free-running mode, it can be set up without any other ICs in the circuit and can display the output binary code with LEDs. In the first part of this experiment, you will test an ADC0804 using the free-running mode and observe the signals. The manufacturer advises that a momentary low on the $\overline{WR}$ line will assure start-up during the first power-up cycle. This is done by a normally open pushbutton switch.

A concern with ADCs is noise and glitches (switching "spikes"). An ADC resolves the input signal in the millivolt and microvolt range. Any noise can have an adverse effect on the conversion accuracy as the internal comparator can respond to the noise. The manufacturer recommends that a 1 μF (or better) tantalum (low inductance) capacitor be connected between the V_{CC} pin and ground. Another concern is assuring that the converter's ground line is not sharing high currents from other parts of the circuit; high currents create unwanted conduction noise that can affect the 0 V level for the converter. Ground currents can be reduced by using fairly high-value current-limiting resistors in series with the LED indicators.

In this experiment, you are introduced to an IC temperature transducer, the LM335 (drawn as an "adjustable" zener diode). The LM335 is internally calibrated for an output of 10 mV per Kelvin degree[1]. Related devices are the LM34 and LM35 designed for measuring positive Fahrenheit and Centigrade temperatures, respectively. At 0° C, the LM335 should have an output of 2730 mV (273 x 10 mV) whereas the LM35 will have a 0 V output (representing 0° C). The LM335 can be used to read either Fahrenheit or Centigrade temperatures at the output of the ADC0804 as you will see. This process will give you experience with controlling the span and offset for a measurement.

Materials Needed:

Resistors: one 1.0 kΩ, eight 1.2 kΩ, one 2.2 kΩ, one 10 kΩ
Potentiometers: one 1 kΩ (ten-turn), two 10 kΩ
Capacitors: one 150 pF, one 0.1 μF, one 1.0 μF (or larger) tantalum
One ADC0804 analog-to-digital converter
One N.O. pushbutton switch
Eight light-emitting diodes (LEDs)
One LM335 IC temperature sensor
One 1N4619 3.0 V zener diode

---

[1] The Kelvin scale uses the same size degree units as the Celsius scale and is based on the coldest possible temperature as 0 K. Equivalent temperatures are 273° larger than those on the Celsius scale.

Procedure:

1. Construct the circuit shown in Figure 35-2. Generally the power supply voltage is set to +5.0 V; however, it is shown here as +5.12 V to make the output steps exactly 20 mV. (5.12 V / 256 steps = 20 mV/ step). The $V_{ref}/2$ voltage is set internally to one-half the supply voltage unless an external reference voltage is connected to this pin. For this test circuit, an external 10 kΩ potentiometer sets the reference. The 3.0 V zener diode is added to ensure this point is fixed regardless of any power supply fluctuations.

   R_2 is a ten-turn potentiometer used to set a precise analog input between 0 and 5.12 V. If you do not have a ten-turn potentiometer, a regular one-turn potentiometer can be substituted (but with reduced precision for setting the voltages).

   C_3 is a 1.0 μF (or larger) tantalum bypass capacitor that should be connected close to the V_{CC} of the ADC0804. If you experience noise problems, try increasing this capacitor.

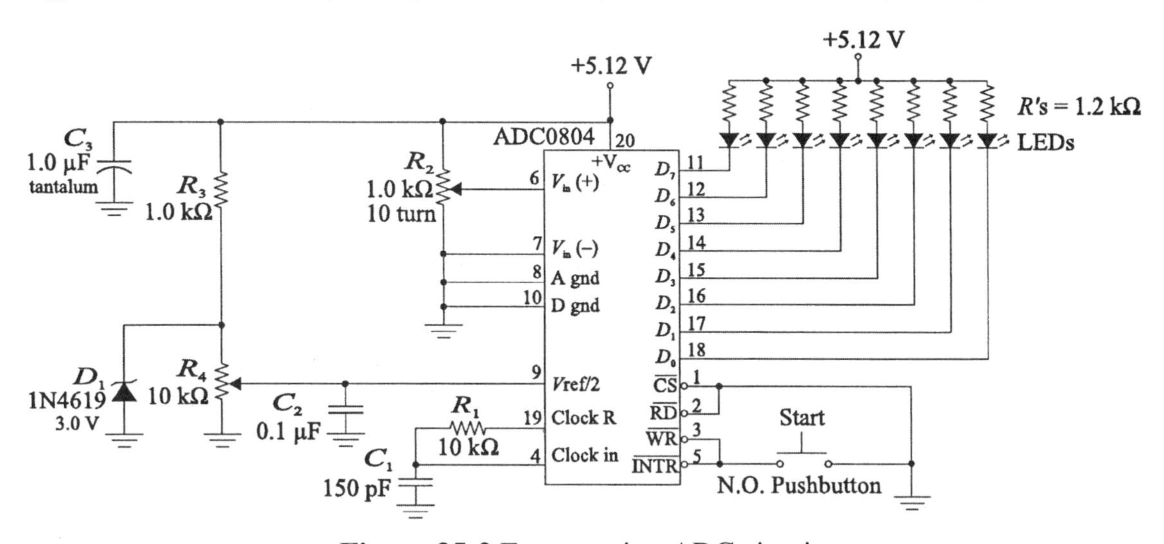

Figure 35-2 Free-running ADC circuit.

2. Adjust the power supply to +5.12 V and momentarily push the Start pushbutton. The ADC should be operating. The internal clock speed is controlled by R_1 and C_1.

   Set the R_2 potentiometer to its lowest resistance setting (0 V) and observe the LEDs. They should all be on. Slowly increase the analog input voltage to maximum. You should observe that the pattern in the LEDs changes as the input voltage is increased until they are all off. Notice that an LED that is *on* represents a binary 0 and an LED that is *off* represents a binary 1. If you do not observe the changing pattern described, troubleshoot the circuit to see that it was wired correctly.

Testing the output

3. This step "fine tunes" the threshold voltages and sets the steps to 20 mV per step, except for the very last step which is 30 mV wide. The adjustments required should be small.

   Adjust R_2 for an input voltage ($V_{in}(+)$) of 5.09 V. This represents 1.5 steps (30 mV) *less* than the full scale of 5.12 V for the last step. At this point, all LEDs except D_0 should be off. (D_0 may be either on or off.) While observing the D_0 LED, readjust the $V_{ref}/2$ voltage using potentiometer R_4 so that the output is *just* changing between 0 and 1. This setting of R_4 should be left for remaining tests.

4. In this step, you will check several steps in the transfer function. The transfer function is a plot of the analog input as a function of the digital output. An ideal transfer function is shown in Figure 35-3 for the first 4 steps (out of a possible 256 steps) for the free-running ADC circuit. Notice that the output ideally switches at points 1/2 between the least significant bit step value.

   Set the R_2 for minimum (0 V). All the LEDs should be on. Adjust R_2 until the D_0 LED just goes out (representing a binary 1). Measure the voltage on the ($V_{in}(+)$) input. This represents the first threshold voltage. Ideally it should be 10 mV, but it may be slightly larger due to current in the resistance of the ground conductor of the protoboard. Record the measured value in the first space of Table 35-1.

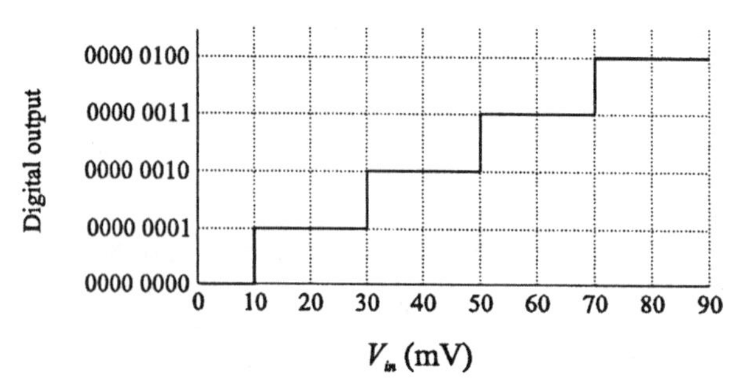

Figure 35-3 Ideal transfer function showing the first four steps for the ADC in Figure 35-1.

Table 35-1

Digital number	Ideal threshold	Measured threshold
0000 0001	10 mV	
0000 0010	30 mV	
0000 0011	50 mV	
0000 0100	70 mV	

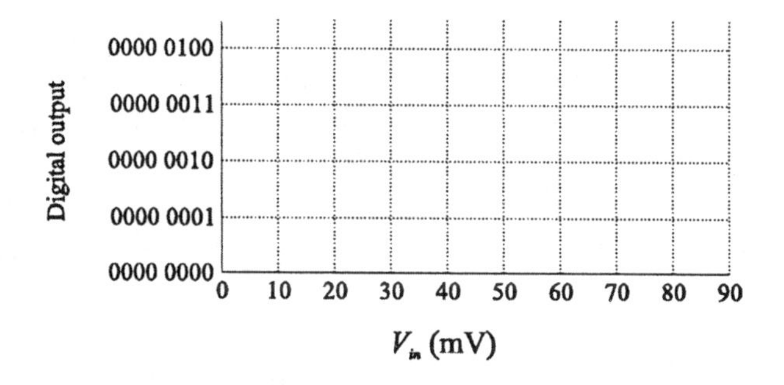

Plot 35-1

5. Raise the input voltage and observe the point where the binary code switches to 0000 0010. Record this threshold in Table 35-1. Repeat for the next two codes in the sequence. Plot the transfer function for the first four steps of your circuit in Plot 35-1.

6. Starting from the least significant position, each binary bit position represents twice the value of the preceding bit. For example, if the least significant bit represents 20 mV, the second position represents 40 mV, the third represents 80 mV and so forth. In this step you will test the input voltage required to turn off a single LED.

   Adjust R_2 carefully for 20 mV. Record the observed binary output in Table 35-2. Remember that an LED that is off is a binary 1.

7. Repeat step 6 for each of the input voltages listed in Table 35-2. Set the input voltage as accurately as possible and record the binary pattern in Table 35-2.

Table 35-2

Input voltage	Binary output
20 mV	
40 mV	
80 mV	
160 mV	
320 mV	
640 mV	
1.28 V	
2.56 V	

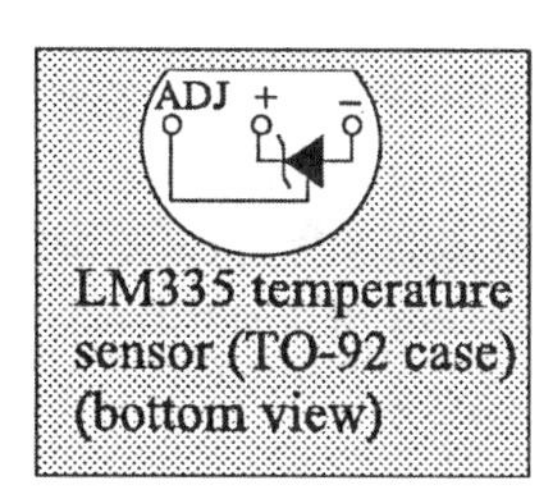

Figure 35-4

Forming a digital thermometer

8. An analog temperature transducer is the LM335, described in the Summary of Theory. The pinout is shown in Figure 35-4. Replace the input circuitry for pins 6 and 7 on the ADC as shown in Figure 35-5 to accommodate the LM335.

   The ADC circuit you have been testing has 20 mV steps. Once V_{CC} is set, these steps are determined by the voltage on the $V_{ref}/2$ input. Since the LM335 has 10 mV/ °C output, change the ADC to 10 mV/step also. You can change the ADC to 10 mV/step by changing the voltage on the $V_{ref}/2$ input to 1.28 V (one-half the desired full-scale range).

   Another adjustment will convert the output of the ADC to represent a Celsius temperature scale. In order to force the output to represent 0° C when the temperature in Kelvin is 273 K, a voltage of 2.73 V is placed on the inverting input (pin 7) of the ADC0804 using the ten-turn potentiometer to set the voltage.

   Can you read the temperature of the room? Try holding the temperature sensor between your fingers. Test the circuit and describe your results:

   ____

   ____

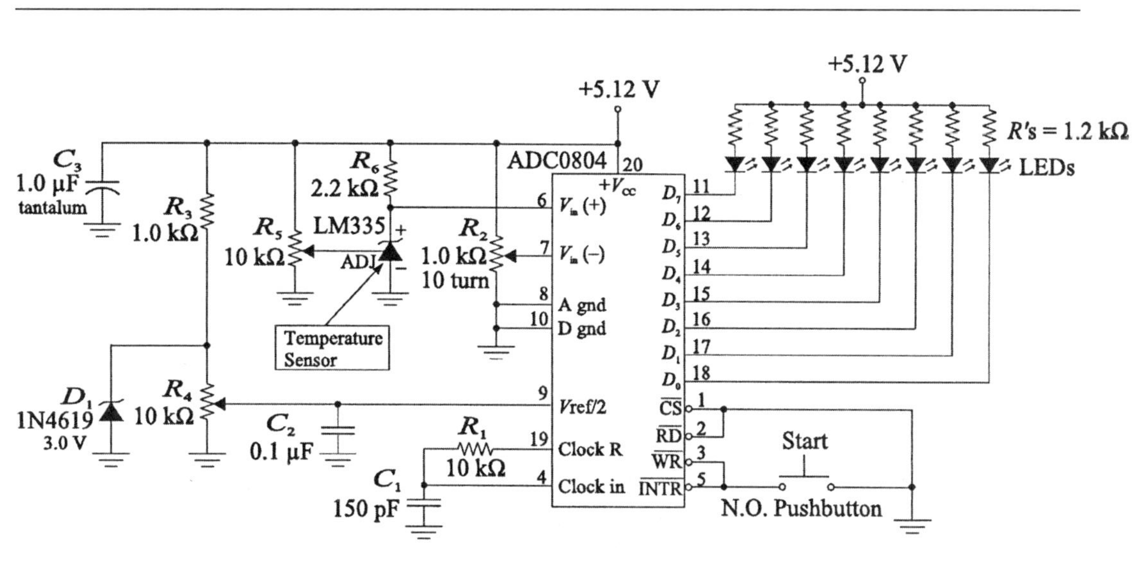

Figure 35-5

Conclusion:

Evaluation and Review Questions:

1. What binary output would you expect if the input to the circuit in Figure 35-2 was 800 mV?

2. If the binary output in Figure 35-2 is 1010 0000, what *input* voltage ($V_{in(+)}$) would you expect to observe?

3. In step 8, you were instructed to change the $V_{ref}/2$ input to 1.28 V. Explain how this setting produces a 10 mV/step output.

4. In step 8, what temperature is represented by an output of 1111 1111?

5. What is the purpose of the $\overline{WR}$ line on the ADC0804?

For Further Investigation:

In step 8, you set the span and offset of the ADC to convert Kelvin temperatures above 273 K from the LM335 to positive Centigrade temperatures at the output of the ADC. By properly setting the span and offset, you can do the same thing for positive Fahrenheit temperatures.

The equation that is used to convert Centigrade temperature to Fahrenheit is:

$$F = \frac{9}{5}C + 32^\circ$$

where:

F = temperature in Fahrenheit
C = temperature in Centigrade

Think about the problem and see if you can figure out what setting of R_2 , R_4, and R_5 will enable the circuit to have a 10 mV/step output for positive Fahrenheit temperatures. Test your settings and summarize your procedures and your results.

36 Transducers

Reading:
Floyd and Buchla, *Analog Fundamentals: A Systems Approach*, Sections 15-3 and 15-4

Objectives:
After performing this experiment, you will be able to:

1. Connect a capacitive displacement measurement system containing a capacitive transducer and a relaxation oscillator.
2. Measure the frequency of a relaxation oscillator as a function of displacement.
3. Construct the transfer curve for the system and use it to determine an unknown number of pages.

Summary of Theory:
Electronic measurement systems use a transducer to convert the physical quantity to be measured into an electrical signal. The complete system includes a transmission path for the electrical signal, signal conditioning, and signal processing to convert the signal into a usable form for the user. Most industrial processes, medical research, and scientific experiments require measurements of physical parameters such as temperature, flow, fluid level, speed, proximity, light intensity, concentration, and the like. The transducer's output can be used directly to control or monitor the process or it can be converted to a digital number and processed by a computer.

There are many types of transducers with a variety of operating principles. To change a physical parameter to one that can be measured by an electronic system requires some electrical parameter to vary. Electrical parameters include change in resistance, capacitance, inductance, magnetic field, voltage, current, or other parameter. In this experiment, the transducer is a simple capacitor constructed from two blank unetched PC boards.

Capacitive transducers are widely used in electronic measuring systems. Recall from your basic dc/ac circuits course that the capacitance between the plates of a capacitor is inversely proportional to the distance of separation. This idea is put to use in measuring pressure, displacement, motion, force, and acceleration and more. A common measuring technique is to use the capacitor to control an oscillator's frequency. The frequency is measured and related back to the original quantity that was being measured.

As discussed in the text, the *transfer function* of a system is the ratio of the output signal to the input. In this experiment, the output signal is a pulse waveform with a variable frequency; the input is the spacing between the plates (represented by sheets of paper) of the capacitive transducer. After taking data for the transducer you have constructed, you will plot its transfer curve.

Materials Needed:
Resistors: two 470 kΩ
One 555 timer
Two blank (unetched) single or double sided PC boards (approximately 8 1/2" x 11")
One hundred sheets of plain paper (8 1/2" x 11")
Small weight (a few hundred grams)
Frequency counter (if available)

For Further Investigation:
A series of 1, 2, and 5 kg masses or other standard weights

Procedure:

1. Construct a capacitive displacement transducer by soldering a wire to each of two blank 8 1/2" x 11" PC boards as shown in Figure 36-1. The size of the boards is not critical but it is a good idea to keep a record of all data in an experiment so you could reconstruct it. Enter the measured size of the board in Table 36-1. Solder wires to opposite sides of each board and place the boards together, forming a sandwich (with no filling). Add a small weight (such as a book) on top to keep the arrangement stable. The boards now form the parallel plates of a capacitor.

Table 36-1

Quantity	Measured Value
PC board size	

2. The 555 timer in Figure 36-1 is connected as an astable oscillator (see Experiment 29). With the transducer plates directly over each other, and no space between the boards, measure the oscillator frequency. You can do this with a frequency counter (if available) or with an oscilloscope. Record the measured frequency in the first space of Table 36-2.

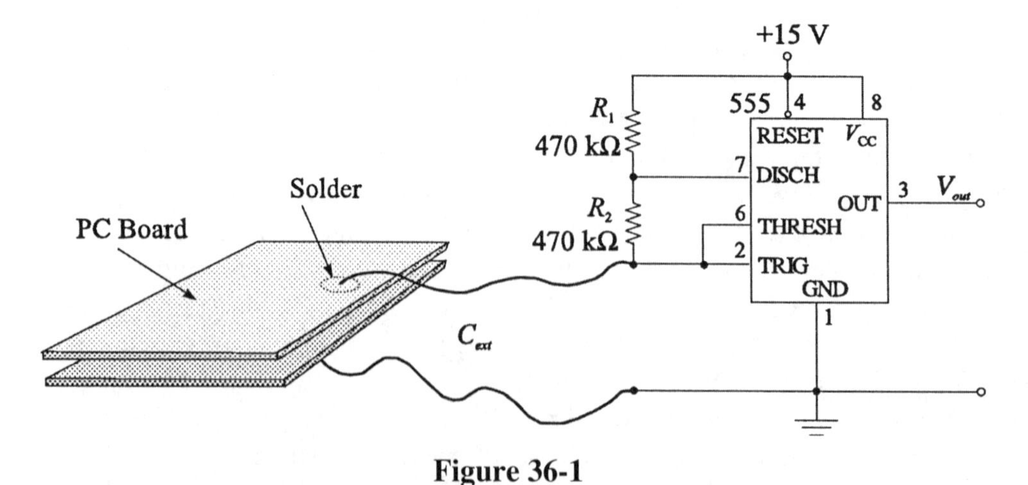

Figure 36-1

3. Place 20 sheets of paper between the plates to act as a uniform spacer; replace the weight on top of the plates. Observe V_{out} on an oscilloscope. Measure the frequency with 20 sheets of paper between the plates. Record the frequency in Table 36-1.

Table 36-2

Number of pages	Measured Frequency
none	
20	
40	
60	
80	
100	

4. Add 20 more pages between the plates; repeat the frequency measurement. Continue in this manner for each stack of paper listed in Table 36-2.

5. Plot points for the transfer function for the experiment on Plot 36-1. Label both axes of your plot. In this experiment, the independent variable (input) is the page count; the dependent variable (output) is the frequency. Draw a smooth "best-fit" line that shows the trend of the data. (It is not necessary to have every point in the line.)

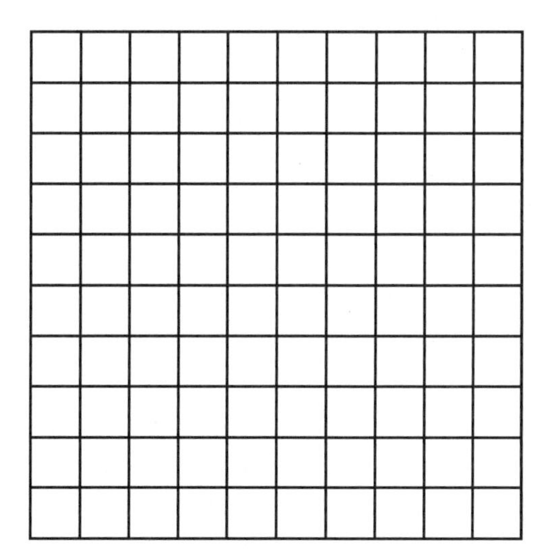

Plot 36-1

6. Interpolation is the process of determining an unknown value by observing known values on either side of a point. From the data you plotted, could you determine how many pieces of paper are in an unknown stack? Try placing (or have a lab partner place) an unknown number of pages (up to 100) between the plates. Measure the frequency. From the measured frequency, predict the number of pages.

   Observations: ____

   ____

Conclusion:

Evaluation and Review Questions:

1. Commercial capacitance sensors tend to have much closer plate spacing than the thickness of the PC board used in this experiment. What advantage would this have?

2. (a) What errors do you think are important in measuring the number of pages?

   (b) Why might the plate alignment or the size of the plates affect the resolution of the displacement measurement?

3. What problems would you expect if you needed to operate the transducer a long way from the oscillator?

4. A consideration for many measurements has to do with the speed of the measurement. How fast do you think the capacitive displacement measurement could respond to a change that a frequency counter would be able to detect? Justify your answer.

5. Suggest a method that you could use to convert the frequency count (as measured on a frequency counter) to a direct reading of the number of sheets of paper between the plates.

For Further Investigation:

Capacitive sensors are used in many different measurements. The capacitive sensor in this experiment can be used to determine the approximate mass of an object. Set up the sensor from Figure 36-1 with 10 pages between the plates and measure the output frequency of the timer with no weight on the plate. Then add a series of known weights (try 1 kg, 2 kg, 5 kg, etc.) and calibrate the sensor. Can you use it to weigh an unknown? Can you find a material that works better between the plates than paper? Does the position of the weight affect the results? Summarize your findings.

Name ______________
Date ______________
Class ______________

37 Measuring Rotational Speed

Reading:
Floyd and Buchla, *Analog Fundamentals: A Systems Approach*, Section 15-4

Objectives:
After performing this experiment, you will be able to:
1. Set up an optical transducer (LED and phototransistor) for making indirect measurements.
2. Using the optical transducer from objective 1, measure the rotational speed (in rpm) of the motor as a function of the applied voltage.

Summary of Theory:
Noncontacting transducers have a great advantage in motion measurement. A common technique for measuring rotational motion is with light sensors (sensing a hole in a floppy disk, for example). In this experiment, you will use a light source and sensor to detect the speed of a motor.

There is a large variety of light sensors, and the spectral responses of the various sensors differ significantly. In addition, there are differences in sensitivity to light, geometric considerations, bandwidth, cost, and ability to operate in different environments. In spite of the great variety, nearly all light sensors can be classed into one of three basic categories: (1) photovoltaic sensors that convert light directly into an emf, (2) photoconductive sensors, which act as light-sensitive resistors, and (3) photoemissive sensors, which contain a light-sensitive cathode that emits electrons when struck by light.

Photodiodes are photoconductive sensors constructed from a *pn* diode junction using nonmetallic or metallic compounds with an overlying light-sensitive layer. When a photon of light passes through the transparent layer, it can be absorbed; the process moves an electron from the valence band into the conduction band, creating an electron-hole pair. If the energy of the photon is high enough, the electron will exceed the bandgap and there is current if the circuit is closed with an external resistor.

Phototransistors are similar to photodiodes but have current gain, typically from 100 to 1000 times that of a photodiode. Base current is provided by photons striking the reverse-biased base-collector junction. Phototransistors are useful where a sensitive sensor with a small active area is needed but, because of nonlinear response and poor temperature characteristics, they are not useful for light measurement applications.

Some very useful transducers use light as their means of transmitting information. An example is a noncontacting motion detector that senses light as it moves. A phototransistor can be arranged to observe light through holes in an encoding disk or to count fringes painted on the surface to be measured. A useful technique for measuring angular speed is to rotate a shutter over a photosensitive element.

Materials Needed:
Resistors: one 1.0 kΩ, one 68 kΩ
One 10 μF capacitor
One small dc motor - 6 V to 18 V rating
Thin cardboard - 3" x 5" index card (or similar)
One NPN phototransistor - MRD300 (or equivalent)
One red LED (TIL228 or equivalent)
Masking tape
2 pieces of #22 or #24 solid wire, 30 cm long

Procedure:

1. In this experiment, you will measure the rotational speed of a rotating motor using a phototransistor and an LED light source. Obtain a small dc motor (6 V to 18 V) and clamp it in a manner shown in Figure 37-1. (*Note:* a ring stand with a buret clamp borrowed from the science dept. works well.)

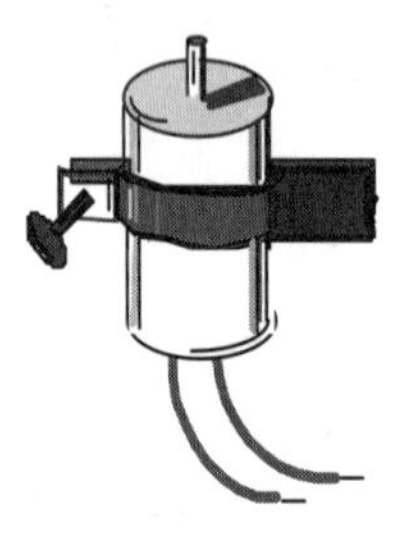

Figure 37-1

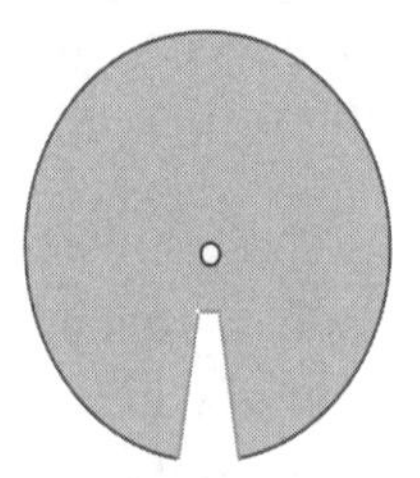

Figure 37-2

2. From light cardboard (such as a 3" x 5" index card), cut a circle with a notch as shown in Figure 37-2. The diameter of the circle needs to be 3 or 4 cm larger than the diameter of the motor. The size of the notch is not critical but will be used to allow light from an LED to strike a phototransistor as the motor turns. (In practice, a hole could be used, but the alignment must be precise.) Fit the cardboard tightly onto the shaft of the motor.

3. Solder 30 cm leads (using #22 or #24 solid wire) onto the collector and emitter of a MRD300 phototransistor and onto the cathode and anode of a red LED. Mount the phototransistor below the disk as shown in Figure 37-3. You may be able to tape it with masking tape to the side of the motor - be careful not to short out the leads. Mount the red LED directly above the phototransistor (mounting method not shown). The leads from the phototransistor and the LED go to a protoboard. Connect the motor leads to a separate dc power supply.

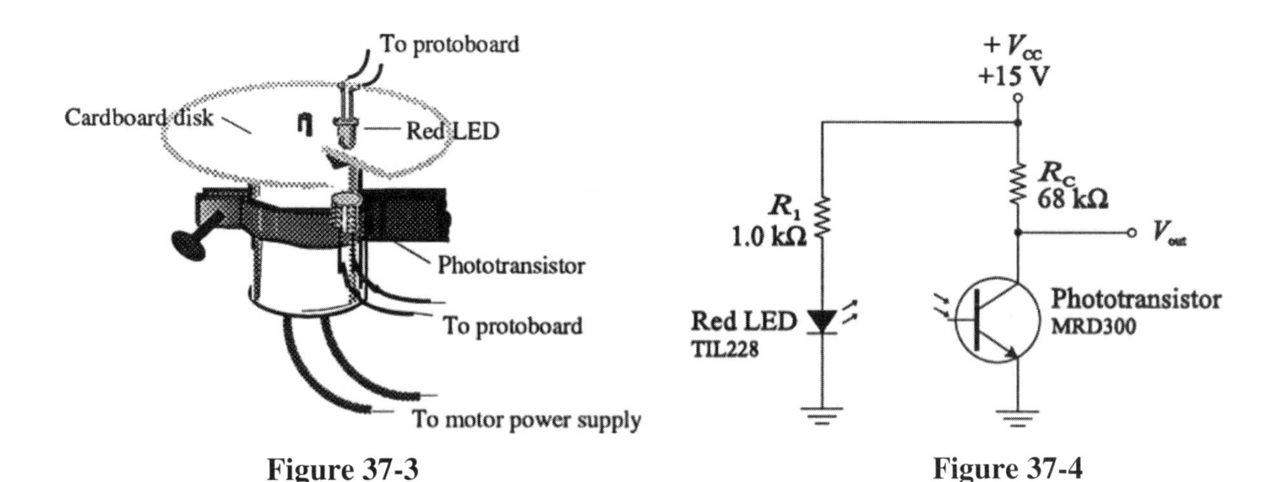

Figure 37-3 **Figure 37-4**

4. Connect the circuit shown in Figure 37-4. Turn on the motor power supply to the rated voltage for the motor and observe V_{out} with your oscilloscope. You should see pulses that are synchronized to the motor speed. You may need to adjust the position of the LED or phototransistor to optimize the signal. This simple system comprises a transducer that converts speed into an electrical signal. (*Note*: some optical couplers are set up for a similar measurement.)

5. Now you can use the transducer you constructed to take data on the speed of the motor as a function of the applied voltage. Set the voltage to 20% of the rated motor voltage. Measure the frequency of the signal. Convert the frequency to revolutions per minute (rpm), a more widely used unit for motor speed. Enter the data in Table 37-1.

6. Adjust the power supply to each of the percentages listed in Table 37-1. Measure the frequency at each setting and convert the reading to rpm. Plot the data for the voltage versus speed in Plot 37-1. Add labels to your plot.

Table 37-1

Motor Voltage	Measured Frequency	Motor rpm
20%		
40%		
60%		
80%		
100%		

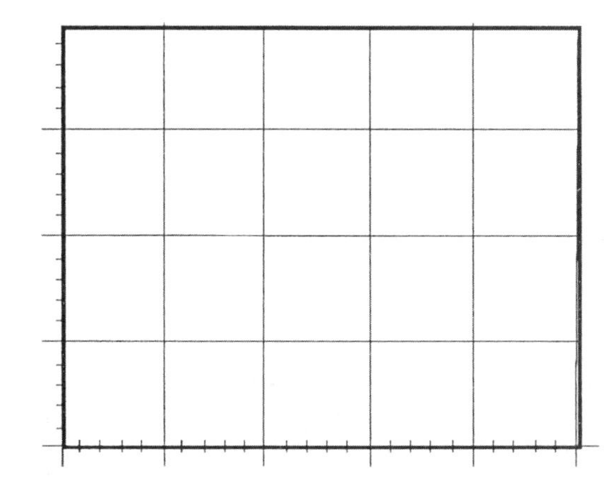

Plot 37-1

Conclusion:

Evaluation and Review Questions:

1. What are some advantages to using an optical system for measuring the angular speed of the motor?

2. Is it important to mount the LED so that the phototransistor "sees" the end of it? Why or why not?

3. Why doesn't the transistor in Figure 37-4 require bias resistors? What supplies base current?

4. If a motor is running at 4000 rpm, what frequency (in Hertz) would you observe at the output of the transistor in this experiment?

5. Review Section 15-4 in the text. How does a tachometer differ from the light detector in this circuit?

For Further Investigation:

Can the phototransistor can be turned on or off with a signal generator? Connect the output of a signal generator to the base of the phototransistor. Vary the dc offset control. What happens? Does the transistor still respond to light? Summarize the answers to these questions in a short report.

38 The SCR

Name ______________
Date ______________
Class ______________

Reading:
Floyd and Buchla, *Analog Fundamentals: A Systems Approach, Section* 15-5

Objectives:
After performing this experiment, you will be able to:

1. Measure the gate trigger voltage and holding current for an SCR.
2. Compare a transistor latch circuit with an SCR.
3. Explain the effect of varying the gate control on the voltage waveforms in an ac SCR circuit.

Summary of Theory:
Thyristors are semiconductor devices consisting of multiple layers of alternating *p* and *n* material. They are bistable devices which use two, three, or four terminals to control either ac or dc. Thyristors are primarily used in power control and switching applications. A variety of geometry and gate arrangements are available, leading to various types of thyristors such as the diac, triac, and silicon-controlled rectifier (SCR). In this experiment, you will investigate an SCR, one of the oldest and most popular of the thyristors. It is a four-layer device and can be represented as equivalent *pnp* and *npn* transistors as shown in Figure 38-1.

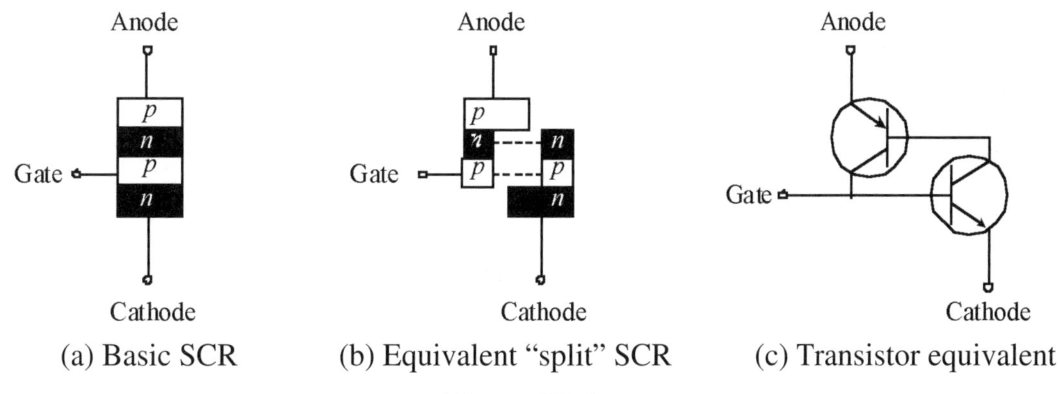

(a) Basic SCR (b) Equivalent "split" SCR (c) Transistor equivalent

Figure 38-1

The SCR is a thyristor which operates as a latching switch controlled by a sensitive gate. If the anode is negative with respect to the cathode, the SCR is reverse-biased and will be off. When the anode is more positive than the cathode, the SCR is forward-biased; but without a gate signal, it remains off. The application of a small positive gate pulse causes the SCR to go rapidly into conduction. Once it begins conduction, control is lost by the gate. Conduction ceases only when the anode current is brought below a value called the *holding current.*

SCRs have specific requirements for proper triggering. A number of special thyristors and other solid-state devices such as the unijunction transistor (UJT) are used for trigger circuits. The primary requirement of any triggering circuit is to provide adequate gate current and voltage at a precise time. The triggering device provides a precise control signal to a thyristor power device. Applications include dc switching, motor control, electronic ignition, battery chargers, and lamp drivers.

Materials Needed:
Resistors: One 160 Ω, two 1.0 kΩ, one 10 kΩ
One 10 kΩ potentiometer
One 0.1 μF capacitor
One LED
One 2N3904 *npn* transistor (or equivalent)
One 2N3906 *pnp* transistor (or equivalent)
One SK3950 SCR (or equivalent)
One 12.6 V power transformer

For Further Investigation:
One photocell (Electronix Express 08GL7516 or equivalent)

Procedure:

1. Measure and record the value of the resistors listed in Table 38-1. R_2 is a 10 kΩ variable resistor, so is not listed.

Table 38-1

Resistor	Listed Value	Measured Value
R_1	1.0 kΩ	
R_3	160 Ω	
R_4	1.0 kΩ	
R_5	10 kΩ	

Table 38-2

Parameter	Transistor Latch	SCR
$V_{AK(off\ state)}$		
$V_{AK(on\ state)}$		
$V_{Gate\ trigger}$		
V_{R4}		
$I_{H(min)}$		

2. Construct the transistor latch shown in Figure 38-2. The purpose of C_1 is to prevent noise from triggering the latch. The switch can be made from a piece of wire. Set R_2 for the maximum resistance and close SW1. The LED should be off. Measure the voltage across the latch shown in Table 38-2 as $V_{AK(off\ state)}$. (V_{AK} refers to the voltage from the anode to cathode.) Enter the measured voltage in Table 38-2 under Transistor Latch.

3. Slowly decrease the resistance of R_2 until the LED comes on. Measure V_{AK} with the LED on (latch closed). Record this value as $V_{AK(on\ state)}$ in Table 38-2. Measure the voltage across R_3. Record this as the gate trigger voltage in Table 38-2.

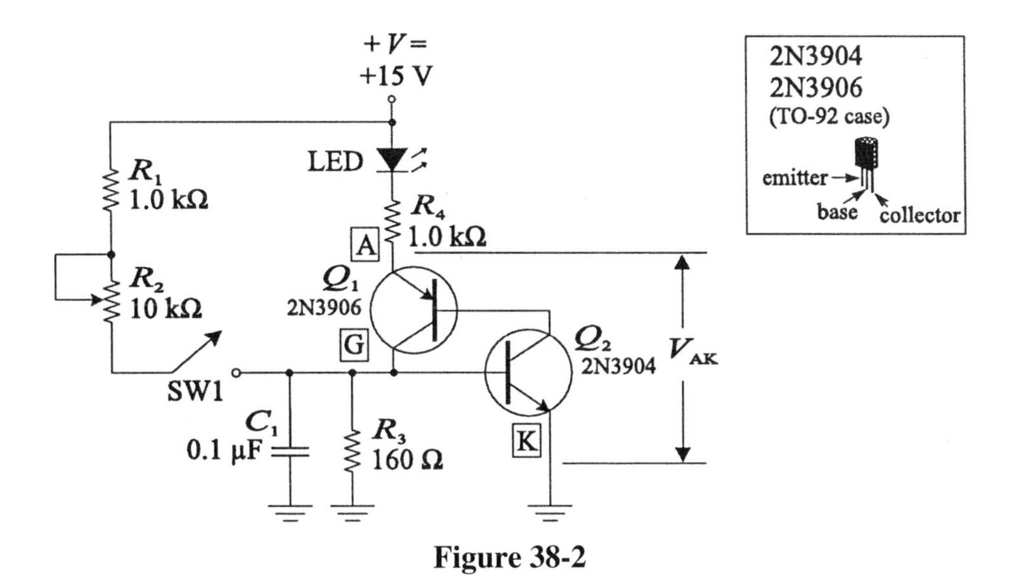

Figure 38-2

4. Open SW1. The LED should stay on because of latching action. Connect a voltmeter across R_4. Monitor the voltage while slowly decreasing the positive supply voltage. Record in Table 38-2, the smallest voltage you can obtain across R_4 with the LED on. Then apply Ohm's law to R_4 to compute the current through R_4. This current is the minimum holding current for the transistor latch. Record this as $I_{H(min)}$ in Table 38-2.

5. Replace the transistor latch with an SCR as shown in Figure 38-3. Repeat steps 2, 3, and 4 for the SCR. Enter the data in Table 38-2 under SCR.

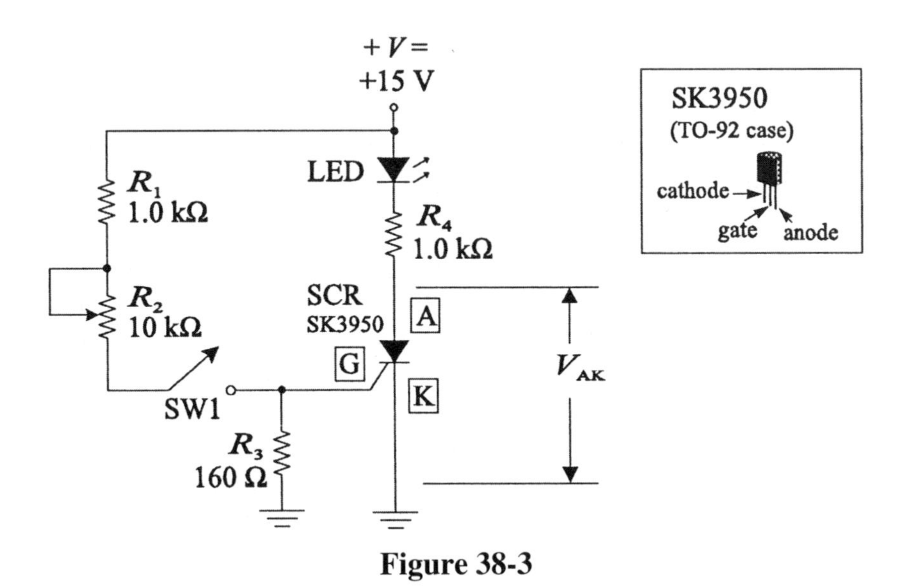

Figure 38-3

6. As you have seen, the only way to turn off an SCR is to drop the conduction to a value below the holding current. A circuit which can do this for dc operation is shown in Figure 38-4. In this circuit, the capacitor is charged to approximately +15 V (the supply voltage). When SW2 is momentarily closed, the capacitor is connected in reverse across the SCR, causing the SCR to drop out of conduction. This is called *capacitor commutation*. Add the commutation circuit to the SCR circuit.

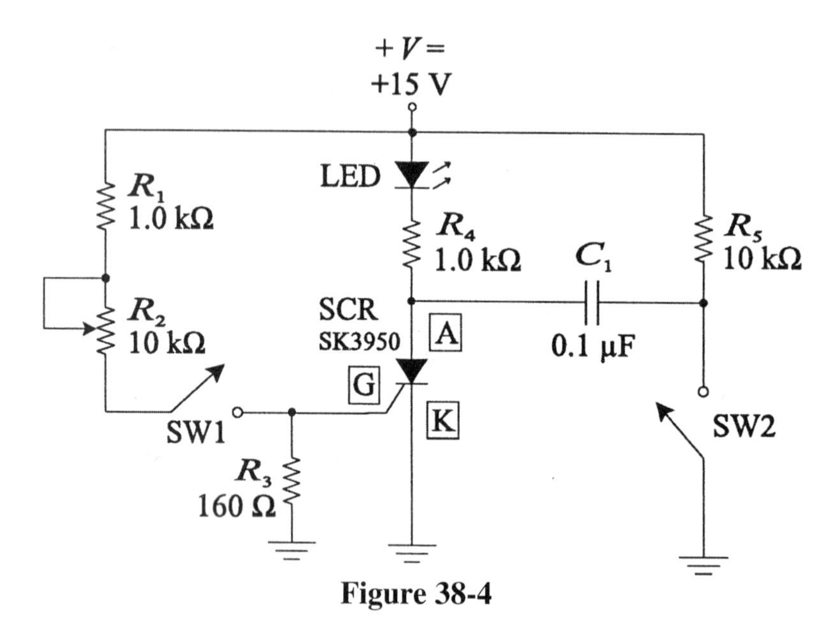

Figure 38-4

7. Test the commutation circuit by <u>momentarily</u> closing SW1 and then <u>momentarily</u> closing SW2. Describe your observations.

____

____

8. **<u>Caution</u>!** In this procedure, you are instructed to connect a low-voltage (12.6 VAC) transformer to the ac line. Be certain that you are using a properly fused and grounded transformer that has no exposed primary leads. Do not touch any connection in the circuit. At no time will you make a measurement on the primary side of the transformer. Have your connections checked by your instructor before applying power to the circuit.

   A typical application of SCRs is in ac circuits such as motor-speed controls. The ac voltage is rectified by the SCR and applied to a dc motor. Control is obtained by triggering the gate during the positive alteration of the ac voltage. The SCR drops out of conduction on each negative half-cycle; therefore, a commutation circuit is unnecessary. Remove the commutation circuit and replace the positive dc supply with a 12.6 V rms voltage from a low-voltage power transformer.[1] Observe the voltage waveform across R_4 by connecting one channel of your oscilloscope on each side of R_4 and measuring the voltage difference between the probes.[2] Compare this waveform with the voltage waveform across the SCR anode to cathode. Vary R_2 and observe the effect on the waveforms. On Plot 38-1, sketch representative waveforms across R_4 and across the SCR. Show the measured voltage on your sketch.

---

[1] A signal generator can be used instead. Set the generator for a 15 V peak signal at 60 Hz.

[2] To find the voltage difference, set both channels to the same VOLTS/DIV setting. Most oscilloscopes will have an ADD control. ADD the channels and INVERT CH 2.

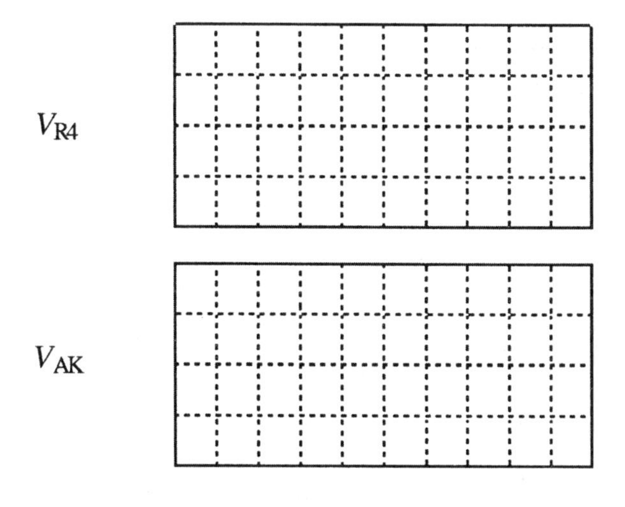

Plot 38-1

Conclusion:

Evaluation and Review Questions:

1. Explain how to turn off a conducting SCR in a dc circuit.

2. What does commutation in an SCR circuit refer to?

3. Explain why a short from the dc supply to the anode of the SCR in Figure 38-3 could cause the SCR to burn out.

4. What symptom would you expect to see if the SCR in the circuit of Figure 38-4 had an anode-to-cathode short?

5. For the circuit of Figure 38-4, what effect on the voltage waveform measured across R_4 would you expect if the holding current for the SCR were higher?

For Further Investigation:

The trigger control circuit of an SCR can be controlled by a light-sensitive detector such as a photocell. Operate the circuit shown in Figure 38-3 from a sinusoidal wave source. Replace R_3 with a photocell (Electronix Express 08GL7516). Then put R_3 back and replace R_2 with the photocell. Summarize your findings in a laboratory report. There are various resistive sensors on the market for temperature, moisture, and so forth. Can you think of other potential applications for this circuit?

Appendix

List of Materials for the Experiments (one of each except where noted)

Resistors:
10 Ω
22 Ω (1/4 W and 2 W)
47 Ω
100 Ω (two)
160 Ω
180 Ω
220 Ω
330 Ω (four)
470 Ω
510 Ω
560 Ω
620 Ω
1.0 kΩ (three)
1.2 kΩ
1.5 kΩ
1.8 kΩ
2.0 kΩ
2.2 kΩ (two)
2.7 kΩ
3.3 kΩ
3.6 kΩ
3.9 kΩ
4.7 kΩ (two)
5.1 kΩ
5.6 kΩ
6.8 kΩ
8.2 kΩ (four)
10 kΩ (five)
15 kΩ
18 kΩ
20 kΩ
22 kΩ (two)
27 kΩ
33 kΩ
47 kΩ (two)
56 kΩ
68 kΩ
82 kΩ
100 kΩ (three)
220 kΩ
330 kΩ
360 kΩ
470 kΩ
1.0 MΩ

Capacitors:
100 pF (five)
1000 pF (three)
2200 pF
0.01 μF (four)
0.1 μF (three)
0.22 μF
1.0 μF (three)
10 μF
47 μF
100 μF (two)
1000 μF (two)

Potentiometers
100 Ω
1.0 kΩ
5.0 kΩ
10 kΩ
100 kΩ

Inductors
2 μH (can be wound from #22 wire)
25 μH
15 mH

Diodes
1N914 signal diode (two)
1N4001 rectifier diodes (four)
1N4733 5 V zener (two)
LED one red, one yellow, one green (TIL228, TIL224, TIL 228)

Transistors and ICs
2N3904 *npn* small-signal transistor (three)
2N3906 *pnp* small-signal transistor (two)
2N5458 *n*-channel JFET (two)
555 timer
741C op-amp (three)
7493A ripple counter
MPF102 *n*-channel JFET
MRD300 phototransistor
MRD500 photodiode
SK3024 *npn* power transistor with heat sink

Miscellaneous
9 V battery
12.6 VAC center-tapped transformer; fused primary
bulb #1869 or 327
CdS photocell - Electronix Express 08GL7516 or equivalent
index card
low impedance microphone (speaker may be substituted)
milliammeter (one 10 mA, one 50 mA)
MV2115 varactor (two)
SK3950 SCR
small 8 Ω speaker
small dc motor 6 to 18 V
optional - frequency counter

Additional materials needed for Further Investigations:
1.0 MHz crystal
7805 or 78L05 regulator
7812 or 78L12 regulator
bulb - #44
heat shrink tubing (for light baffle) that fits over phototransistor
masking tape
milliammeter 0 - 50 mA
optional - transistor curve tracer
potentiometer (500 Ω)

List of Materials for the Experiments (by experiment number)

Experiment 1: Signal Sources and Amplifier Characteristics

One 2N3904 *npn* transistor (or equivalent)
Resistors: one 1.0 kΩ, one 1.8 kΩ, one 3.9 kΩ, one 10 kΩ
One 10 μF capacitor
One 1.0 kΩ potentiometer

Experiment 2: The Diode Characteristic

Resistors: one 330 Ω, one 1.0 MΩ
One signal diode (1N914 or equivalent)

Experiment 3: Rectifier Circuits

Resistors: two 2.2 kΩ resistors
One 12.6 V ac center-tapped transformer with fused line cord
Four diodes 1N4001 (or equivalent)
One 100 μF capacitor
For Further Investigation:
One 0.01 μF capacitor
One 7812 or 78L12 regulator

Experiment 4: Diode Limiting and Clamping Circuits

Resistors: two 10 kΩ, one 47 kΩ
Two signal diodes: 1N914 (or equivalent)
One 47 μF capacitor
For Further Investigation:
Three 1.0 kΩ resistors

Experiment 5: Special-Purpose Diodes

Resistors: one 1.0 kΩ, one 1.0 MΩ
One 10 kΩ potentiometer
Two MV2115 varactors
One MRD500 photodiode
One 15 mH inductor.
One small transformer with a 12.6 V secondary
Three LEDs, one red, one yellow, one green
Light source (bright lamp or flashlight)
For Further Investigation:
One MRD300 phototransistor
Resistors: one 510 Ω, one 330 kΩ
Masking tape
Heat shrink tubing (for light baffle) that fits over phototransistor

Experiment 6: Bipolar Junction Transistor Characteristics

Resistors: One 100 Ω resistor, one 33 kΩ resistor
One 2N3904 *npn* transistor (or equivalent)
For Further Investigation:
Option 1: Transistor curve tracer
Option 2: One rectifier diode (1N4001 or equivalent)
One small power transformer with a 12.6 V ac output

Experiment 7: Bipolar Transistor Biasing

Resistors (one of each): 470 Ω, 2.0 kΩ, 6.8 kΩ, 33 kΩ, 360 kΩ, 1.0 MΩ
Three small signal *npn* transistors, (2N3904 or equivalent)
For Further Investigation:
Resistors: one 3.6 kΩ, one 100 kΩ

Experiment 8: The Common-Emitter Amplifier

Resistors: one 100 Ω, one 330 Ω, two 1.0 kΩ, one 4.7 kΩ, two 10 kΩ
Capacitors: two 1.0 μF, one 47 μF
One 10 kΩ potentiometer
One 2N3904 *npn* transistor (or equivalent)

Experiment 9: The Common-Collector Amplifier

Resistors: two 1.0 kΩ, one 10 kΩ, one 33 kΩ
Capacitors: one 1.0 μF, one 10 μF
One 10 kΩ potentiometer
One 2N3906 *pnp* transistor (or equivalent)
For Further Investigation:
One 2N3906 *pnp* transistor (or equivalent)
Three 330 Ω resistors

Experiment 10: Transistor Switches

Resistors: one 330 Ω, one 1.0 kΩ, two 10 kΩ
One 10 kΩ potentiometer
Two small signal *npn* transistors (2N3904 or equivalent)
One LED

Experiment 11: 11 JFET Characteristics

One 100 Ω resistor
One 10 kΩ resistor
One 2N5458 *n* channel JFET (or equivalent). (Save the JFET for Experiment 12.)
One LED
One milliammeter 0-10 mA range

Experiment 12: JFET Biasing

Resistors: one 2.2 kΩ, one 330 kΩ, one 1.0 MΩ, two values to be determined by student
One 1.0 kΩ potentiometer
One 2N5458 JFET
One milliammeter 0 - 10 mA
For Further Investigation:
one additional 2N5458 JFET

Experiment 13: FET Amplifiers

Resistors: two 1.0 kΩ, one 3.3 kΩ, one 10 kΩ, one 100 kΩ, one 1.0 MΩ, one to be determined by student
Two 2N5458 *n*-channel JFETs (or equivalent)
Capacitors: one 0.1 μF, one 1.0 μF, one 10 μF
For Further Investigation:
One 1 kΩ potentiometer
One 620 Ω resistor

Experiment 14: JFET Applications

Resistors (one of each): 180 Ω, 2.7 kΩ, 3.9 kΩ, 5.1 kΩ, 27 kΩ, 56 kΩ, 1.0 MΩ
Capacitors (one of each): 0.1 μF, 1.0 μF, 10 μF
Transistors: one 2N3904 *npn* transistor, two 2N5458 *n*-channel JFETs (or equivalent)
One LED
One milliammeter 0 - 10 mA

Experiment 15: Multistage Amplifiers

One 2N3904 *npn* transistor (or equivalent)
One 2N3906 *pnp* transistor (or equivalent)
Capacitors: two 0.1 μF, three 1.0 μF
Resistors: one of each: 220 Ω, 1.0 kΩ, 2.0 kΩ, 4.7 kΩ, 6.8 kΩ, 10 kΩ, 33 kΩ, 47 kΩ
Resistors: two of each 22 kΩ, 100 kΩ, 330 kΩ
One 100 kΩ potentiometer
For Further Investigation:
One MPF102 *n*-channel JFET
One 1N914 signal diode (or equivalent)
Resistors: one 100 Ω, one 220 kΩ
Capacitors: one additional 1.0 μF capacitor

Experiment 16: Class A Power Amplifiers

Resistors: one 22 Ω (2 W), one 100 Ω, one 560 Ω, one 4.7 kΩ, two 10 kΩ, one 22 kΩ, one 56 kΩ
Capacitors: one 0.22 μF, one 1.0 μF, one 10 μF, two 100 μF
Transistors: two 2N3904, one SK3024 (or equivalent) with heat sink
One small 8 Ω speaker
For Further Investigation:
One additional 2N3904 transistor
One low resistance microphone (a small speaker can be substituted)
Resistors as specified by student

Experiment 17: Class B Push-Pull Amplifiers

Resistors: one 330 Ω, one 2.7 kΩ, two 10 kΩ, one 68 kΩ
One 1.0 μF capacitor
Transistors: one 2N3906 *pnp*, two 2N3904 *npn* (or equivalent)
Two 1N914 diodes (or equivalent)
One 5 kΩ potentiometer
For Further Investigation:
One 15 kΩ resistor, one additional resistor to be determined by student

Experiment 18: The Differential Amplifier

Resistors: two 100 Ω, two 10 kΩ, one 33 kΩ, two 100 kΩ
Capacitors: two 10 μF
Transistors: two 2N3904 (or equivalent)
For Further Investigation:
One additional 2N3904 transistor
Two additional 10 kΩ resistors and one 4.7 kΩ resistor

Experiment 19: Op-Amp Characteristics

Resistors: two 100 Ω, two 10 kΩ, two 100 kΩ, one 1.0 MΩ
Two 1.0 μF capacitors
One 741C op-amp
For Further Investigation:
One 10 Ω resistor

Experiment 20: Linear Op-Amp Circuits

Resistors: two 1.0 kΩ, one 10 kΩ, one 470 kΩ, one 1.0 MΩ
Two 1.0 μF capacitors
One 741C op-amp
For Further Investigation:
One 1.0 kΩ potentiometer, one 100 kΩ resistor, assorted resistors to test

Experiment 21: Op-Amp Frequency Response

Resistors: one 620 Ω, two 1.0 kΩ, one 2.0 kΩ, one 3.3 kΩ, one 10 kΩ, one 18 kΩ, one 100 kΩ
Two 1.0 μF capacitors
One 741C op-amp
For Further Investigation:
One 100 Ω resistor

Experiment 22: Comparators and the Schmitt Trigger

Resistors: one 100 kΩ
Two 1.0 μF capacitors
One 10 kΩ potentiometer
One 741C op-amp
For Further Investigation:
One additional 100 kΩ resistor
One 0.1 μF capacitor

Experiment 23: Summing Amplifiers

Resistors: one 3.9 kΩ, one 5.1 kΩ, four 10 kΩ, one 20 kΩ
Capacitors: two 1.0 μF
Two signal diodes, 1N914 (or equivalent)
Two op-amps: LM741C
One 7493A 4-bit ripple counter
For Further Investigation:
Resistors: two 4.7 kΩ, three 100 kΩ
Capacitors: one 0.01 μF, one 0.1 μF

Experiment 24: The Integrator and Differentiator

Resistors: two 1.0 kΩ, four 10 kΩ, two 22 kΩ, one 330 kΩ
Capacitors: one 2200 pF, one 0.01 μF, two 1.0 μF
Three 741C op-amps
One 1.0 kΩ potentiometer
Two LEDs (one red, one green)
For Further Investigation:
Resistors: two 4.7 kΩ, two 100 kΩ
One 0.1 μF capacitor

Experiment 25: Low-Pass and High-Pass Active Filters

Resistors: one 1.5 kΩ, four 8.2 kΩ, two 10 kΩ, one 22 kΩ, one 27 kΩ
Capacitors: four 0.01 μF, four 1.0 μF
Two 741C op-amps
For Further Investigation:
One additional 741C op-amp and components to be specified by student

Experiment 26: State-Variable Band-Pass Filter

Resistors: three 1.0 kΩ, three 10 kΩ, one 100 kΩ
Capacitors: two 0.1 μF, two 1.0 μF
Three 741C op-amps
For Further Investigation:
Resistors: two additional 10 kΩ, one 82 kΩ
One 10 kΩ potentiometer
One 0.01 μF capacitor

Experiment 27: The Wien Bridge Oscillator

Resistors: one 1.0 kΩ, three 10 kΩ
Capacitors: two 0.01 μF, three 1.0 μF
Two 1N914 signal diodes (or equivalent)
One 741C op-amp
One 2N5458 *n*-channel JFET transistor (or equivalent)
One 10 kΩ potentiometer
For Further Investigation:
One type #1869 or type #327 bulb

Experiment 28: The Colpitts and Hartley Oscillator

One 2N3904 *npn* transistor (or equivalent)
One 100 Ω potentiometer
Resistors: one 1.0 kΩ, one 2.7 kΩ, one 3.3 kΩ, one 10 kΩ
Capacitors: two 1000 pF, one 0.01 μF, three 0.1 μF
Inductors: one 2 μH (can be wound quickly from #22 wire), one 25 μH
For Further Investigation:
One 1.0 MHz crystal
One 2N5458 n-channel JFET
One 1.0 MΩ resistor

Experiment 29: The 555 Timer

Resistors: one 1.0 kΩ, one 9.1 kΩ, one 10 kΩ, one 22 kΩ, one 1.0 MΩ
Capacitors: one 0.01 μF, one 0.1 μF
One 555 timer IC
One 1N914 diode (or equivalent)
One CdS photocell (Radio-Shack 276-116) or equivalent
For Further Investigation:
One 100 kΩ potentiometer
One 0.1 μF capacitor

Experiment 30: The Zener Regulator

Resistors: one 220 Ω (1/2 W), one 1.0 kΩ, one 2.2 kΩ
One 1.0 kΩ potentiometer
One center-tapped transformer with fused primary, 12.6 V ac
One 5 V zener, (1N4733 or equivalent)
For Further Investigation:
Second 5 V zener, (1N4733 or equivalent)

Experiment 31: Voltage Regulators

Resistors: four 330 Ω resistors, one 1.0 kΩ, one 1.2 kΩ, one 2.7 kΩ
One 1.0 kΩ potentiometer
Transistors: one 2N3904 *npn* transistor, one SK3024 *npn* power transistor
Diodes: one 1N4733 5 V zener diode, four 1N4001 rectifier diodes
One 1000 μF capacitor
For Further Investigation:
One 7805 or 78L05 regulator
One 220 Ω resistor
One LED
One 0 – 50 mA ammeter

Experiment 32: The Instrumentation Amplifier

Resistors: one 22 Ω, one 470 Ω, two 1.0 kΩ, one 8.2 kΩ, five 10 kΩ, three 100 kΩ
Capacitors: one 0.01 μF, two 1.0 μF
Three LM741C op-amps
One 555 timer
One small 9 V battery
One 5 kΩ potentiometer
30 cm twisted-pair wire
For Further Investigation:
One CdS cell – Electronix Express 08GL7516 or equivalent
One 10 kΩ potentiometer
One 0.01 μF capacitor

Experiment 33: Log and Antilog Amplifiers

Resistors: two 100 kΩ
Capacitors: one 0.01 μF, two 1.0 μF
Two 1N914 signal diodes (or equivalent)
Two 741C op-amps
Two 2N3904 *npn* transistors (or equivalent) (*Note:* You will obtain best results if ßs match).
For Further Investigation:
One additional 741C op-amp
Resistors: one 5.1 kΩ, two 10 kΩ
Potentiometers: one 500 Ω, one 10 kΩ

Experiment 34: IF Amplifiers

3rd stage IF transformer (20 kΩ primary to 5 kΩ secondary) Mouser 42IF303 (or equivalent)
Resistors: one 220 Ω, one 470 Ω, one 4.7 kΩ, two 10 kΩ, one 56 kΩ
Capacitors: two 0.1 μF
Transistors: two 2N3904 *npn* (or equivalent)
Frequency counter (if available) (An oscilloscope can substitute.)
For Further Investigation:
Resistors: one 100 Ω, one 270 Ω, one 1.0 kΩ, one 2.0 kΩ

Experiment 35: The ADC804 Analog to Digital Converter

Resistors: one 1.0 kΩ, eight 1.2 kΩ, one 2.2 kΩ, one 10 kΩ
Potentiometers: one 1 kΩ (ten-turn), two 10 kΩ
Capacitors: one 150 pF, one 0.1 μF, one 1.0 μF (or larger) tantalum
One ADC0804 analog-to-digital converter
One N.O. pushbutton switch
Eight light-emitting diodes (LEDs)
One LM335 IC temperature sensor
One 1N4619 3.0 V zener diode

Experiment 36: Transducers

Resistors: two 470 kΩ
One 555 timer
Two blank (unetched) single or double sided PC boards (approximately 8 1/2" x 11")
One hundred sheets of plain paper (8 1/2" x 11")
Small weight (a few hundred grams)
Frequency counter (if available)
For Further Investigation:
A series of 1, 2, and 5 kg masses or other standard weights

Experiment 37: Measuring Rotational Speed

Resistors: one 1.0 kΩ, one 68 kΩ
One 10 μF capacitor
One small dc motor - 6 V to 18 V rating
Thin cardboard - 3" x 5" index card (or similar)
One NPN phototransistor - MRD300 (or equivalent)
One red LED (TIL228 or equivalent)
Masking tape
2 pieces of #22 or #24 solid wire, 30 cm long

Experiment 38: The SCR

Resistors: One 160 Ω, two 1.0 kΩ, one 10 kΩ
One 10 kΩ potentiometer
One 0.1 μF capacitor
One LED
One 2N3904 *npn* transistor (or equivalent)
One 2N3906 *pnp* transistor (or equivalent)
One SK3950 SCR (or equivalent)
One 12.6 V power transformer
For Further Investigation:
One photocell – Electronix Express 08GL7516 or equivalent